Modern Science and the Capriciousness of Nature

Also by Karl Rogers

ON THE METAPHYSICS OF EXPERIMENTAL PHYSICS

Modern Science and the Capriciousness of Nature

Karl Rogers

First published 2006 by
PALGRAVE MACMILLAN
Houndmills, Basingstoke, Hampshire RG21 6XS and
175 Fifth Avenue, New York, N.Y. 10010
Companies and representatives throughout the world

PALGRAVE MACMILLAN is the global academic imprint of the Palgrave Macmillan division of St. Martin's Press, LLC and of Palgrave Macmillan Ltd. Macmillan® is a registered trademark in the United States, United Kingdom and other countries. Palgrave is a registered trademark in the European Union and other countries.

ISBN 978-1-349-54222-2   ISBN 978-0-230-62519-8 (eBook)
DOI 10.1057/9780230625198

This book is printed on paper suitable for recycling and made from fully managed and sustained forest sources.

A catalogue record for this book is available from the British Library.

Library of Congress Cataloging-in-Publication Data
Rogers, Karl, 1967–
    Modern science and the capriciousness of nature / Karl Rogers.
        p.   cm.
    Includes bibliographical references and index.

    1. Science—Philosophy.   2. Philosophy of nature.   3. Technology.
    I. Title.
    Q175.R537 2006
    501—dc22   2006043636

10   9   8   7   6   5   4   3   2   1
15   14   13   12   11   10   09   08   07   06

For Caitlin

For serious thinkers today and for growing numbers of politically active persons as well, the progressivist view of history seems neither true nor appealing. The dubious metaphysical underpinnings of the faith [in inevitability of progress through science and technology] are now embarrassingly obvious. Mounting signs of the social and environmental problems that 'progress' generates have eroded the conviction that there is a necessary link between technological advance and the well-being of the Earth and its people. Insofar as the aspirations of democratic societies still hinge upon the outward progressivist mythology, those aspirations are in deep trouble.

Langdon Winner, Democracy in the Technological Society (1992)

Every passing hour brings the Solar System forty-three thousand miles closer to Globular Cluster M13 in Hercules – and still there are some misfits who insist that there is no such thing as progress.

Kurt Vonnegut, The Sirens of Titan (1959)

Contents

This page intentionally left blank

1

The Capriciousness of Nature

The recent earthquakes and tsunamis in Asia and the hurricanes in the Caribbean and the United States remind us of the indiscriminate destructive power of Nature. Clearly the poor people of affected countries are the least protected and suffer the most during the aftermath of such natural catastrophes, but the destructive power of Nature is indifferent to class, race, creed, wealth, or politics. We live in a world in which hunger, poverty, disease, squalor, fear, misery, and premature death are widespread consequences of natural catastrophes. If we also consider floods, epidemics, blizzards, lightning strikes, landslides, avalanches, tornadoes, and droughts, then we can readily recognise that the destructive power of Nature seems arbitrary. Babies are all too often born with horrendous deformities, clearly through no fault of their own, and old age frequently brings the pain and indignity of decay, senility, and infirmity, regardless of how the person lived their life. Death ultimately comes to us all regardless of status, wealth, or character. A physically fit athlete can be struck down in his or her prime by a hereditary heart defect, while watched by an overweight and aged spectator who has survived his or her third heart attack after a lifetime's indulgence and inactivity. We seemingly struggle, in the face of numerous arbitrary and destructive events, to live within a hostile or indifferent natural world, which can strike us down without warning or reason. The capriciousness of Nature reminds us that human existence and our capacity for well-being are fragile and vulnerable, and this is something that we often find very frightening and dreadful. Even when we gaze upon the natural world with awe and wonder, while marvelling at its splendour and beauty, we still, at that moment, share in such praise, being the being that is capable of recognising this in Nature. We take a share in the incredibleness of the natural world, as we stand, seemingly alone in our appreciation, gazing upon something that we presume has no consciousness of what it is for us. Of course, the experience of the Otherness of Nature is often a sense of the extent that the natural world is beyond human comprehension and succinct expression. There is often an intuitive sense of something *je ne sais quoi*, something mysterious beyond our intellectual reach and grasp, in our experience of the natural world. We can recognise, respect, and celebrate this aspect of Nature. For millennia, philosophers and mystics have celebrated this aspect of Otherness, while arguing that human beings are part of Nature and that there is considerable continuity and kinship between human beings (other) animals, and all living beings. By focussing on the wonder, beauty, and bounty of the natural world, they have argued in favour of a holistic, ethical, and harmonious approach to living within the world. However, while I have considerable agreements with these philosophers and I accept that it is true that the natural world is full with wonder, beauty, and bounty, in this book, I shall turn to the dark side, if you will forgive the expression, in order to explore and discuss the human response to the capricious destructiveness of Nature. I do not wish to imply that Nature is malignant, but the human experience of Nature as Otherness is also an experience of Nature as an aloof and indifferent source of destructive power. That is not something for which we take a share of the credit. That is the sense that we are insignificant to the extent that the extinction of us all would pass unnoticed. There would be no tears. There will not even be a sigh of relief. When Nature is understood as something aloof and indifferent to human life, spontaneity in Nature is itself something that is represented as being synonymous with chaos, as uncontrollable, unpredictable, and threatening. Nature is represented as something primordial and the mindless source of all instability and change, in the face of which human action and toil becomes pointless struggle. Everything that we are, we dream of, we work towards, could all be swept away, without warning, without reason, without justice, at any time. The Otherness of Nature looms over us and imposes a perpetual sense of vulnerability, futility, and dread.

One response to this sense is that of the so-called "primitive and polytheistic religions" (such as animism and pantheism), often made in terms of supplication, ritual, and sacrifice, as attempts to appease Nature and the gods. Another is the theological response presented by the monotheistic religions, which are often made in terms of elliptical justifications and appeals to faith, or somewhat *ad hoc* explanations in terms of divine retribution for sinful human behaviour. Religion is often a way of consoling the human sense of dread, fear, and ill ease about the natural world by explaining it in terms of some divine purpose or plan.[1] The capriciousness of Nature has been represented as providing human beings with valuable lessons for self-improvement, as a punishment for a "fall from grace", or as a result of human foolishness.[2] On this argument, the failure of human beings to see this world as "the best of all possible worlds" is a consequence of human ignorance and limited perspective about the divine will or plan. However, the existence of natural disasters is often presented as something that discredits the idea of "divine providence". The assertion that such disasters represent a punishment for sinful behaviour is quite hard to reconcile with the fact that many "sinful" adults escape the disaster, while innocent children (and animals) die either during the disaster or in its aftermath. Many pregnant mothers also die, taking their unborn children with them, and some people die while trying to save others. It also presupposes that "sinfulness" is localised at the site of the disaster, rather than spread globally in more or less equal measure. Hence, it is quite hard to accept that "divine retribution" should take the form of a natural disaster, rather than a more precise and unambiguous form (such as an accurately aimed lightning bolt). It seems that even our religions and most profound spiritual beliefs are hard stretched to explain and console us about the seemingly unjust and arbitrary cruelties that Nature imposes upon human life. The existence of natural disasters makes it quite impossible to reconcile the idea that Nature is well ordered and rational with the idea that human beings are somehow important in the scheme of things. However, it may well be the case that, as mere mortals, we simply cannot understand the place for natural disasters in the divine plan. Perhaps John Stuart Mill was correct not to abandon the idea of an overall design in Nature, but nonetheless argue that any designer of such a world as we experience cannot be all powerful and all good, for nature is "too clumsily made and capriciously governed".[3] The sheer volume of pain, suffering, violence, and cruelty within the natural world makes it difficult to claim that such a seemingly brutal, cruel, and purposeless Nature expresses the love and goodness of an omnipotent God.[4]

Animal life inherently involves feeding upon other living beings. Injury, harm, pain, violence, destruction, decay, and death are an inherent part of the natural world. This world is not a utopian paradise of pleasure and ease, but, instead, is a place in which arbitrary struggle and suffering seem to be inherent to ongoing life. Many geologists and paleobiologists have presented evidence to support the theory that there have been a series of mass extinctions caused by catastrophic events throughout the history of the Earth.[5] Even though we can describe the way that many forms of life take advantage of natural catastrophes and mass extinctions, we should take care not to mistake opportunism for "natural economy" or "natural balance". Nature apparently squanders individuals and species, without respect for their fate, and there is no reason to believe that this is done with any overall consideration for the whole.[6] Whether individuals or species are able to take advantage of such events seems to be more readily characterised as a matter of accident than it is of balance.[7] According to the theory of evolution, random mutations provide the variations from which natural selection and diversity emerges. It is seemingly a double-edged sword. The source of all variation brings with it wonder, beauty, bounty, disability, suffering, and inborn fatal flaws. It seemingly bestows these randomly without any regard for justice or purpose. As Charles Darwin argued, even though the prolific process of evolution is one in which natural selection operates upon random incremental variations with the net result being the survival of some species and the extinction of others, random variations in individuals are as equally likely to be disadvantageous as advantageous.[8] A surviving species finds the world to be beneficial to its survival, simply because it has survived, while, at the same time, remains constantly threatened by any changes in the world. As Frances Crick put it, organisms evolve their structures through happenstance, and biology does not possess elegance or any overarching logic.[9] Or to use Stephen Gould's examples, things such as the panda's thumb and orchids are simply products of trial and error.[10] Perhaps human existence is only a matter of chance, merely an accident. If this is true then we are existentially estranged in a world that is indifferent to our existence.[11] If this is not the case then there are two complaints we could make against Nature: first, there is Nature's capriciousness in bestowing bounty upon some and inflicting destruction upon others, through no fault nor merit of their own. When seen from the human viewpoint, with certain human expectations, this strikes us as unfair. Second, one could say that Nature is ultimately unjust to the degree to which it is indifferent to human need, often being harshest to the most vulnerable, such as children, the disabled, and the elderly, when it is Nature that is the source of their vulnerability and need. As T.H. Huxley quipped, Nature might well stand condemned before the tribunal of human ethics.[12] Of course, it is one thing to describe Nature as indifferent to human interests and well-being, but it is quite another to condemn it as immoral or evil. One needs to distinguish between bad and evil – the former is an aesthetic judgement on the quality of the experience, made in relation to our lives, expectations, and purposes, whereas the latter is a moral judgement on the motivation or intention that caused the event. Natural catastrophes are evidently bad for the beings that suffer them, but it does not follow from this that they are evil or the consequence of evil. It has even been argued, by social evolutionists since Herbert Spencer, that the struggle for survival is itself a good that acts as a filter to remove those incapable of surviving in order to create space for those better suited for this evolving world.[13] However, we need to be aware that our use of terms such as "hostility", "indifference", "aloofness", "good", "bad", or "capriciousness" does imply a purposeful will or attitude, even if we deny any consciously malicious intent on the part of Nature. The very social nature of our language projects human intention and motivation, in the form of metaphors, analogies, and personifications, to non-human entities; we characterise the behaviour of natural beings in accordance with the descriptions and prescriptions that we use to describe and prescribe our own behaviour. It is a consequence of the fact that we are purposeful beings, with a vested interest in our own well-being, that we can characterise events in terms of aesthetic judgements, such as good or bad, fortunate or unfortunate, advantageous or disadvantageous.[14]

It is in this sense that I describe Nature as being capricious, without intending to imply or deny any intentionality to Nature. Whether or not Nature is conscious and has intentions is a question that is beyond the scope of this book. Nor does this book present a moral judgement against Nature based on value-laden interpretations of scientific results and theories. The purpose of this book is to explain how we use modern science in order to cope with the capriciousness of Nature by constructing a technological society to surround us and protect us, removing that sense of powerlessness, vulnerability, and dread. I shall argue that modern science is a response to this fear and is the consequence of the desire for certainty and power in a world that seems chaotic, aloof, indifferent, and often hostile to human life. It is a drive to transform the sense of the Otherness of Nature into a something that we have a victory over by constructing an artificial world within which we are free and protected from its capricious aspect. This book generalises from the arguments presented in my previous book, *On the Metaphysics of Experimental Physics*, and explores the implications of those arguments in greater depth, critically examining and questioning the rationality of the presuppositions and aims of modern science.[15] When the natural world seems capricious in relation to the human urge for rational order – as a reflection of human fear and limitation – the modern scientific response aims to help us come to terms with our fragility by using scientific knowledge and technology to confront Nature. This confrontation is premised upon a societal gamble that is emergent from the human struggle against and within a world that often seems beyond human control and comprehension. The gamble is that modern science will make the world become a better place because it will become more intelligible to human beings and we will become freer by becoming more powerful. My argument is that modern science has, from its onset, been aimed at creating the means to liberate our fellow human beings of the arbitrary capriciousness of Nature and the limits, vulnerability, ignorance, and mortality that it imposes upon us in our "natural state of animality".[16] The societal gamble was premised upon dreams of earthly paradise and the desire to prevail over "the brute nature of things". Technological innovation is central to modern science and is premised upon the ontological and moral argument that human freedom is enhanced by the knowledge of natural laws and mechanisms. Scientific truth is equated with the good by an epistemology that equates rationality with increased technical knowledge and the novel powers associated with it. The societal gamble implicitly presupposes the rationality and goodness of the project of constructing a technological society to replace the natural world with a better artificial one. Yet still the capriciousness of Nature emerges in the form of every new strain of disease and weakness of the body discovered by science. It is present within the operation of technologies as machines break down, fatigue creeps in, materials deteriorate and decay, unforeseeable side effects and consequences abound, and our technologies become new sources of illness, disease, and death. The Otherness of Nature looms and lurks in the imperfections of the technological society, again reminding us of the futility of our efforts, while still motivating us to make further efforts to perfect the technological society. The very technologies that we invented to give us control and mastery over Nature, by discovering and turning its own forces against each other, becomes something beyond our control and alien to us, perhaps making us even more vulnerable than we previously were, as we become increasingly dependent on those technologies. This is the ironic tragedy of the technological society.[17] Moreover, even though technological innovation is driven to transforming the natural world into a better place for human beings, often these interventions have disturbed the natural world, leading to localised instability, which has lead to unexpected natural disasters. Technologies have created problems, such as overpopulation, pollution, and environmental degradation, which lead to further natural disasters and an increasing dependency on technology.

Many of the deaths caused by natural disasters are a consequence of bad decisions on the part or humans (i.e. living in areas prone to earthquakes or building houses out of inappropriate materials), as well as by social inequalities in access to resources and technologies (which forces people to live in risky areas or build their houses out of poor-quality materials). On this account, the onus upon human beings is to learn how to live in harmony with the natural order by using scientific knowledge to create a better social order.[18] The development of this social order, through the innovation and dissemination of science and technology, is one that entails a vision of the ideal society and the human good life based on technoscientific knowledge and power.

Natural science and modern technology

Scientific realism presupposes that natural laws and mechanisms, which govern the phenomenal appearance of the natural world, transcend and pre-exist our efforts to discover and theorise about them. This realism presupposes that any failure of correspondence between our theories and objective reality is not likely to benefit us, but it may do us harm.[19] Scientific theory and practice is considered tested when it is shown to have practical value for human life by solving practical problems, leading to the innovation of novel prototypes, and the further development of scientific activity. Human beings are represented as developing science as part of fulfilling our inborn potential to use our evolved intelligence to survive in the natural world, being in confrontation with the limitations imposed by Nature, by utilising our discoveries of natural laws and mechanisms to innovate new technologies to overcome these limitations. Our ability to struggle against and oppose Nature is represented as being a product of a naturally evolved capacity that actually comes from Nature. Even though the human tendency to personify and describe Nature as morally reprehensible, evil, irrational, or capriciously destructive runs counter to the modern scientific conception of Nature as mechanistic and impersonal, this psychological, existential dread motivates the scientific enterprise. The idea of an inborn capacity to struggle against Nature is one that reflects the aesthetic and instrumental value of scientific knowledge for human life, and, as I shall argue in this book, when combined with this existential dread of the power of Nature, human beings are psychologically driven to construct a technological society to optimise scientific control over the instrumental value of Nature and remove its capricious power over our lives. Nature is transformed into a rationalised source of instrumentality for technological innovation and progress, providing human beings with the power to harness and domesticate Nature, which is ontologically reduced to a set of instrumental powers at the disposal of machines, which are themselves only given meaning from their function within the human lifeworld of purposeful action. On such a view, the natural world is portrayed as an unenchanted material world, governed by natural laws and mechanisms, which can only have instrumental value in relation to human minds.[20] Objective answers are taken to be the remit of the natural sciences, but questions of aesthetics, ethics, and morality are taken to be questions of value and, consequently, without the postulation of a divine mind, are solely human questions that have no objective answers. However, while it is not the task of the natural sciences to raise moral questions or solve moral problems, it is evident that the transformation of Nature into something that is ontologically mechanistic equates the true with the instrumental, which in turn represents truth as an instrumental value for human well-being. As I shall argue in this book, it is this representation that supports the moral obligation of the Enlightenment, which demanded that all scientific research should improve the human condition by providing human beings with increased technological power and control over our material conditions and limitations; science should aid human survival and improve the world. This equation between truth and value underwrites the scientific realist claim that science is tested and has been proved to be successful through its self-evident practical successes.

Value-free science does not mean that science should be something to which no value can be assigned. The fact that biological science has value in providing medical science with means to improve our heath and protect us from disease is not sufficient to demonstrate the value-ladeness of that science. Even identifying core values to the scientific enterprise, such as objectivity, rigorousness, thoroughness, honesty, impartiality, excellence, and so on, is also not sufficient. If value-free science presupposes to show us the objective facts, irrespective of whether they are valuable or to whom they might be valuable, then any individual scientist may be said to be working in the spirit of value-neutrality. What is necessary to demonstrate that the science in question is value-laden is to show that it uses the core values of the society from which it is emergent in the construction of its methodology. Whilst it may well be the case that any individual working scientist may well be indifferent to the particular results, benefits, or risks of his or her research, the research itself would be value-laden if its methodology had already presupposed in its precepts, assumptions, and application, the answer to evaluative questions regarding relevance or importance. If the methodology that an individual uses is value-laden then any "objective facts" that it discovers will also be value-laden. So whilst the individual scientist may well refuse to make any value-judgements about his or her work, the results of the work itself (and not just its possibility) will be the consequence of prior value-judgements (and not necessarily those of the individual scientist). As I argued in *On the Metaphysics of Experimental Physics* and shall discuss further in the next chapter, according to scientific realism, only real mechanisms work in the world and, hence, theoretically postulated mechanisms are experimentally tested through their successful implementation in the innovation of novel machines. The scientific realist representation of the stratification of the sciences as being based on the discovery of ontological depth is dependent on the stratification of the process of technological innovation in accordance with theoretical representations of how those machines work.[21] The natural sciences are based on the methodological reduction of natural philosophy to being the process of construction novel machines and testing theoretical interpretations of machine performances in ongoing technological innovation. Anything outside of this technological framework cannot be established as even being worthy of scientific investigation. Truth is reduced to being that which is disclosed through machine performances in relation to theoretical representations of the underlying mechanisms in operation in order to make those performances possible. The methodology of modern experimental sciences such as physics are inherently technological, within an ongoing process of the technological innovation and modelling of machine prototypes, through a process of mapping out and explaining the interactions between human interventions and machine performances. The methodology of the experimental sciences is premised upon a conceptual equation between truth and technological power. Hence, instrumental value is taken to be a methodological precondition of the establishment of theoretical claims and scientific methods, demonstrating a correspondence with objective reality, and therefore the scientific methodology is value-laden.

Of course, modern scientists have made amazing discoveries (such as penicillin) and brought new powers into the world (such as radio and nuclear power). Science has provided many great benefits to the world, such as communication devices, medicines, medical techniques, light and heating, water purification and irrigation processes, transportation, equipment to help people overcome disabilities, improved weather prediction, and so on. Scientists have simplified our understanding of the world by unifying and explaining disparate phenomena (such as the variation of species of animals and plants, or the motion of the Moon and projectiles on Earth), providing us with an intelligible world-picture. Scientists are able to observe tendencies (such as global warming) and correlations (such as increases in cases of leukaemia in children near nuclear power stations). Scientific instruments have allowed us to discover new and exciting things that were previously hidden from view (such as microbes) that subsequently explained phenomena in the world (such as diseases). Scientists engage in real, interesting, and informative ways with worldly phenomena, as well as bringing new phenomena into existence, which change our lifeworld. This state of affairs is obviously the case. However, this is not the important issue. The important issue is whether and how scientists understand entities and processes within the world. How do scientists explain and model phenomena? What does that tell us about how scientists interact with things in the world? Given the power of the devices produced by experimental physics (i.e. nuclear power, radio communications, and lasers), any misunderstanding or lack of control is something of considerable concern. This is also worrying in the biological sciences, which use genetic modification and manipulation as a technique for understanding the biochemical mechanisms of life itself. It is of paramount importance whether sciences such as physics and genetics are capable of achieving objective knowledge and actually predicting the consequences of their actions. As I shall argue in the next chapter, experimental scientists understand and model things in the world in technological terms, and human beings have come to understand and interact with the natural world as if it were mechanical. Experimental sciences such as physics, chemistry, and genetics have achieved the status of "the natural sciences" because science and technology have been "internally" connected through a realist interpretation of the mathematical science of mechanics. Once we accept that this value-laden interpretation of technology is necessary for the development of modern natural sciences, such as physics, chemistry, and genetics, then we need to examine how the possibilities, structures, and limitations of technology conditions and specifies the content and structure of scientific theories and research. Technology not only is a condition for the possibility of scientific knowledge, but it also mediates and shapes the form of that knowledge. It is not just the case that technology prepares and filters physical reality, by means of specific selective and manipulative operations, which it certainly does, but, more importantly, technology mediates the way that we represent physical reality and how we can engage with it. The very conception of how scientific objectivity could be realised and tested is itself mediated by technology. Moreover, the technological innovation of new powers and the search for theoretical truths are inseparable within the historical development of the experimental sciences. The concept of "causal power" is understood in terms of the realisation and exercising of mechanisms and is central to both the pragmatic and theoretical development of experimental science, as well as the scientific representation of "change" and "process" in the natural world, while demonstration of manipulative and productive powers is taken to be the test and corroboration of scientific theories and causal accounts. Thus, from the onset, the manipulation of natural entities and processes, in order to exhibit the human capacity to possess power over Nature by using technology, has been central to the whole scientific enterprise *qua* natural science. This does not imply that the phenomena investigated by experimental science are only understood in terms of the practical purposes to which they are put, but it does imply that such phenomena are only understood in terms of their capacity to be put to practical use. Technology mediates how we understand the causality by which such phenomena come into being, how we can investigate them, and how we can interact with them. Natural properties are reduced to a set of fundamental representations that are understood and explained in terms of a set of machine performances that are represented as the consequences of the operation of natural mechanisms, corroborated by their further use in the explanation and inscription of novel machine performances in the laboratory, and hence the elaboration of the natural phenomena under investigation is limited to their aspects that are amenable to technological innovation, manipulation, and reproduction. The theoretical understanding of these phenomena is only developed in terms of the technological framework in which they are investigated.

Once scientific causal accounts are given in terms of the mechanisms disclosed through the innovation and modelling of machine performances, then the modern scientist tacitly reintroduces teleology by reconfiguring purpose in the guise of functionality. This abstracts and objectifies the means–ends relationships disclosed through technological innovation and naturalises the instrumental value inherent to the methodology of modern science. As I argued in *On the Metaphysics of Experimental Physics*, this is implicit and essential to the epistemology and methodology of the experimental physics. As I shall argue in the next chapter, this applies to experimental sciences such as genetics, biochemistry, chemistry, and medicine as well, in so far as those sciences are also based on laboratory experiments, rather than observations of natural processes in their natural conditions. From the onset of modern science, the notion of natural order was understood in technological terms, in terms of human methods and material practices, as revealed in the scientific discovery of natural mechanisms and laws through experimentation. The scientific experience of Nature is made meaningful as a series of interactions and events, but when the theoretical understanding of these interactions is tested within the laboratory, then that test is presented as a series of comparisons of models of machine functionality with those of machine performances. It is a comparison of theoretical expectations with experiences within the same technological framework. This is also apparent in the so-called "observational sciences" such as evolutionary biology and ecology where the existence and behaviour of any organism is understood and explained in terms of functions in mechanistic relations with other organisms, to the extent that values themselves can be seen as a function of an evolutionary process that selects in favour of drives towards survival and efficiency. Natural selection is represented as a genetic sorting mechanism: a feedback mechanism between random variation and environmental change.[22] Ecology represents Nature as the ontological consequence and site of a self-reproducing conjunction of events through the exercise and interaction of natural mechanisms in accordance with natural laws. The idea of an ecosystem is based on the notion of a good or bad trait of an animal, plant, or organ being normalised in terms of values and purposes – represented in terms of functions – emergent in relation to the whole system. In James Lovelock's Gaia Hypothesis, for example, Earth is represented as a self-regulating system in which the climate and chemical composition of the planet is coupled to the metabolism of the organisms that inhabit it. The Gaia hypothesis is infused with machinic imagery, but not that of the foundry press or mechanical clock, but rather with that of the self-regulating steam governor or the feedback loops of cybernetic systems.[23] In this sense, the sciences of evolutionary biology and ecology are both value-laden and lead to the development and refinement of a world-picture in which human beings are to be situated within a system of mechanistic actions and reactions naturally selected on the basis of their instrumental value within the whole system. The experimental sciences of psychology and neuroscience are stratified refinements of this world-picture, through further technological innovation, in order to represent the mind as being the product of a system of functions operating upon sensory inputs, using the cybernetic system or the computer as the dominant explanatory trope.[24] The metaphor of the machine situates human beings within the scientific world-picture as being the animal that was capable of manipulating natural mechanisms – through the refinement of our inborn cognitive capacities – in order to release natural powers and possibilities through technological innovation. Technology is naturalised within the scientific world-picture as being the rational implementation of natural mechanisms in practical activity.

This is even more evident when we do not have direct perception acquaintance with the natural entities we are investigating. The innovation of novel technologies to make observations also requires the innovation of novel forms of communication and representation, not only to show others how to make those observations but also to ensure that the observers are all talking about the same thing. Innovation of language use (including visualisation and mathematical techniques) must always proceed against a background of previously successful language use, even when it transforms that background considerably. Thus the innovation of novel observational techniques and instruments is always preconditioned in socially mediated terms, and our experience of Nature is inextricably intertwined with the means by which we understand and interact with Nature. Nature may well be everything that is objectively "out there", so to speak, but the problem is that how human beings understand and interact with Nature is governed by value-laden fundamental representations and concepts. Modern science aims to represent Nature as a homogeneous and isotropic unity of interacting systems that can be predicted and guide the directions of technological innovation. One of the most widely accepted commonplaces is that science is the only proper instrument with which to investigate Nature and that it has progressively expelled metaphysics from its fields of inquiry. Consequently, whenever scientists qualify a statement as being "metaphysical", it invariably means that such a statement must be rejected as an unscientific speculation. Supposedly, any statement or hypothesis about the structure of reality is considered metaphysical if it is based upon a synthetic argument constructed from premises that cannot be empirically tested. However, as I argued in *On the Metaphysics of Experimental Physics*, Newtonian mechanics as well as all the subsequent branches of experimental physics (including optics, thermodynamics, electromagnetism, solid-state physics, and quantum mechanics) are based on the operational metaphysical precepts of mechanical realism. Both natural and technological phenomena are represented as sharing a unitary origin in the same causal principles, coming into being through the sequential exercise of natural mechanisms. The only distinction between natural and technological phenomena is that the latter require human intervention to come into being whilst the former do not. The realisation of any mechanism is governed by natural laws, and, consequently, the performance of any machine is governed by natural laws. The mathematical descriptions of machine performances are mathematical descriptions of natural laws. These metaphysical precepts are necessary in order to epistemologically connect the particular experiences of the "closed system" of the laboratory experiment with the "open system" of the natural world, which transcends and pre-exists the experiment and is a precondition for its intelligibility. Otherwise the artificial practices of experimentation cannot be represented as being part of natural science. Such metaphysical concepts are necessary for the transformation of science, as something merely intuitive and discursive, into a technoscience that "proves" its proximity to truth in terms of technological innovation and practicality. Hence, modern experimental physics utilises metaphors between natural phenomena and machines when it questions how Nature works.[25] How do birds fly? How does the Sun shine? Modern physics answers these questions by making an artificial device fly or shine and representing this device as a model of the natural phenomenon. The technical knowledge of how to make the device is represented as the knowledge of how the natural phenomenon occurred. Once our natural sciences were based on mechanical realist metaphysics, then the knowledge of how to live with Nature was produced by representing our own abilities to intervene in the world and innovate novel technologies as being the realisation and exercising of natural mechanisms in material practices. Once we recognise the centrality of these metaphysical precepts to the construction of the explanation of the empirical facts about the natural world from the artificial practices of experimentation, then we can readily appreciate that the positivistic claim that the content of scientific knowledge is confined to empirical statements divorced from metaphysics is simply false. Given that in the vast majority of cases of experimental research in physics involves entities that we have no direct experience of, such as electromagnetic fields, subatomic particles, or energy states, but are disclosed through the interpretation of machine performances, then even the facts of measurement are determined in relation to a causal account of how the measuring apparatus works.

From time to time, modern sciences may well undergo paradigmatic crises and changes, with incommensurable meanings between paradigms; these epochs of transformation can be situated within a historical continuity, as being radical developments and transformations of the same science.[26] The meaning of quantities such as "mass" made by classical and quantum physicist may well be sufficiently distinct to lead to incommensurable measurements and observations, but both physicists would agree on the value of mathematical modelling, making measurements, and how experiments should be performed. Hence, Aristotelian and Galilean physics are incommensurable at a methodological and metaphysical level, both radically disagreeing about what constitutes science and reality, whilst modern classical and quantum mechanics agree on what constitutes science and are only incommensurable at an interpretational level about the nature of reality. Both classical and quantum physics are connected, through a shared history of technological practices and innovation, as refinements of the same world-picture, and hence quantum physics can be represented as progressive through technological stratification. The innovation of new machines and causal accounts is taken to disclose a deeper stratum of reality. Whilst physicists and philosophers may well argue indefinitely about the meaning of the double-slit experiment, it is the invention of devices such as the laser and tunnelling electron microscope that provide quantum mechanics with its progressive value. It is this shared world-picture embodied in the history of technological innovation that is a precondition for the transference of meanings between classical and quantum mechanics; the choice between quantum and classical physics is not based on a comparison between observations, but is based upon the enhancement of technological possibilities and powers that emerge from the innovation of new machines. Of course, the world-picture is perpetually open to refinement, in accordance with new discoveries, models, and theories, but its basic mechanical realist foundation and trajectory has remained unchanged. Thus the world-picture continues to situate and orientate individuals within a vision of the technological evolution of their lifeworld. It provides the means by which individuals can represent their actions and practices as being in contact with a shared, objective reality, whilst also providing the means to interpret their experiences in accordance with scientific theories, models, and knowledge. Modern scientists positivistically look to the methodology by which they understand phenomena, as being neutral, because they inherit and embody the methodology of their predecessors, while representing that inheritance as being progressive. This inheritance and embodiment provides epistemological, ontological, and moral meanings to whole research projects and particular actions within those projects by situating them within a technological framework. This technological framework allows the subjective experiences and intentions of the individual to be represented as having shared and objective meaning and value. This is only possible because of the equation between power and truth that is central to the mechanical realist foundation of the world-picture, and consequently, both to classical and quantum physics. It is due to the perpetual need to test the products of science by implementing them in the ongoing process of technological innovation that the scientist's never-ending refinement of truth is perpetually placed at the disposal of technology for no other purpose than the extension and development of further technological innovation.

Modern science situates us within a world-picture that we can grasp, whilst leaving that picture open to future refinement. As Martin Heidegger argued, in his essay *The Age of the World Picture*, a world-picture (*Weltbild*) is central to modern scientific conception of the world.[27] "The world" is set in place before us, as a representation, and all that belongs to it and stands together in it is conceived as a system, in such a way that we are acquainted with it as something that we are equipped and prepared to deal with. In this context, the word "picture" is used in the sense of the colloquial expression "to get the picture", to capture the way that we grasp the matter in question, and not in the sense of a copy or imitation. In this way, the successes of modern science in providing us with new technologies, powers, and experiences is the counterpart of the process of making those successes intelligible in terms of a perpetually refined and stratified mechanistic model of the world. As I shall argue in Chapter 3, the world is conceived, visualised, and interacted with in terms of the extent to which it is available for technological innovation and at the disposal of labour. Our satisfaction in the truth (or approximate truth) of this world-picture is secured and enhanced by the sense of liberation and power as new technologies make new experiences and powers possible. As a consequence of this, the positivistic notion of value-free or value-neutral conception of science is not only based upon an imaginary science, but it completely fails to address the fact that any scientific truth claim needs to be instrumentally valuable in order to be considered to be a candidate for truth. Even at the level of a speculation or hypothesis, it is essential that some technical connection with a problem, novel experience, or anomaly is made to the extent that, in order to qualify as a scientific hypothesis, some novel technique must be derived from that hypothesis and technologically implemented within the ongoing activities of real scientific work. A value-free or neutral hypothesis or theory would be independent from the practical context of technological innovation within the ongoing activities of experimentation, and, as a consequence, would be mere speculation and scientifically valueless because it would be untestable.

The neutrality of working scientists is itself a social ambition that is only realised by the implementation of novel techniques and machine prototypes within the ongoing innovation of the technological framework that it itself presupposed to be a neutral, rational application of our natural intelligence. This sense of rationality is itself dependent upon the presupposition of the metaphysics of mechanical realism. Each refinement of the world-picture is represented as progress towards an idealised future world in which science enlightens and liberates human beings from their own ignorance and folly. Where we suffer from the consequences of natural events, the capriciousness of Nature is represented as a consequence of the disorganisation and absence of human knowledge and agency. Our disharmonious existence within the natural world is taken to be a consequence of our incompleteness and imperfection, both of which are to be overcome by the rational use of science and technology. This attempt to situate us within a world-picture, as isolated and alienated beings, combines radical individualism with the mechanical realist conception of Man as the being who discovers his objective conditions and possibilities through labour, by overcoming material limitations. Technology is represented as objectively based on natural mechanisms and laws, and, consequently, Nature is disclosed through the technological processes and procedures of the natural sciences as being objective, value-neutral, and devoid of purposes and ends, while being disclosed through purposeful human activity in confrontation with Nature.

Once we examine the function of purposes and ends in initiating and transforming the trajectories and meaning of research within science, then it becomes increasingly difficult to locate the origin of purposes and ends within the human subject, or, to put it another way, it becomes increasingly important to examine the role of modern science in the construction of society. Whilst the societal gamble results from an aesthetic judgement upon the quality of human life without technological aids, a judgement against the aesthetic quality of the "natural state" of human beings, it does not necessarily result from a moral judgement against Nature. The moral judgement that underwrites the societal gamble is a moral imperative to participate in the ongoing construction of the technological society based on the presumption that the technological society will ultimately be better for human beings than our "natural state". It is the product of a social complex of aesthetic and existential choices in favour of a scientific and technological society. This presupposition makes a virtue out of an adherence to the axiomatic value of the impartial pursuit of scientific truth because it is presupposed that scientific truth is a good that will help us to universally transcend this "natural state" and achieve a good life free from the tyranny and capriciousness of Nature. Thus the societal gamble represents a rejection of stoic fatalism, acceptance, and virtue, in favour of an ongoing process of creating the "best of all possible worlds" – an artificial paradise on Earth. An expression of the societal gamble is most apparent in Richard Dawkins' *The Selfish Gene*, when he argued that, through genetic science, we enhance our Promethean ability to simulate and predict the future, and gain power over the tyranny of our selfish genes.[28] However, it is not only the case that modern science provides us with the technological means to transform the material conditions of our world in order to enhance our productive capabilities and liberate us from the limitations of our "natural state"; it also involves the construction of an aesthetic experience of the world within which enhanced power and intelligibility are central to the conception of rationality and progress. The aesthetic judgement in favour of a scientific understanding of Nature can be readily inferred from Steven Weinberg's final sentence in *The First Three Minutes*, when he wrote that fundamental science has existential value because it "is one of the very few things that lifts human existence a little above the level of farce, and gives it some of the grace of tragedy".[29] Furthermore, as Carl Sagan argued, modern science also promises to provide certainty and understanding in a world that otherwise seems meaningless and indifferent, as a liberating pursuit against superstition and false consciousness.[30] Having world-picture at one's disposal allows one to orientate oneself within a network of relations between entities, processes, functions, and events, in such a way as to provide experiences and expectations with rational, scientific value. This connects human motility and intentionality within a whole complex of modal possibilities, modelled in terms of causal sequences and mechanisms, and it is this connection that permits the embodied practitioner to acquire and utilise technical knowledge as scientific knowledge. However, the world-picture also provides universal conceptions of both historical and evolutional goals and expectations for a scientifically empowered humanity, and it is this aspect of the world-picture that allows individuals to represent their actions as participating in the struggle for the promised goods and truths of the modern world. The world-picture not only contains presuppositions regarding the nature of reality but it also contains presuppositions of the nature of the human good life and visions of the ideal world. As a consequence of this, having a world-picture at one's disposal allows one to orientate oneself in such a way as to represent one's participation in scientific and technological activities as having universal aesthetic and moral value, no matter how small that participation may be.

The world-picture is one that is established by the extent that it offers us a collection of stories that transform "a barren wilderness of white noise into a habitable world".[31] Of course, any individual engaged in diverse educational and cultural interactions learns these stories in a fragmentary and incomplete way, and, as such, often the individual tacitly embodies the world-picture in seemingly disconnected practical activities, representations, discourses, and narratives. Over the course of a lifetime, the individual tacitly acquires a vast sedimented stock of scientific and technical knowledge, alongside with its meaning and value, often without any reflection upon that knowledge, how it was acquired, and how it is to be applied in a consistent way. This tacit knowledge is more often than not acquired without any reflection or understanding of the vast history of human effort required for it to be educationally and culturally available. It is within this history that the connections and meanings of the world-picture have emerged and been refined, whilst its trajectories and legacies frame, shape, empower, and limit contemporary research and imagination, and, as such, the world-picture transcends the individual for whom it is implicit. It is the embodiment of these historical connections and meanings that allows the individual to participate within this historical enterprise, and also empowers the intuition of the individual to see the world-picture for oneself, as if one had discovered it anew. Thus the individual is able to see and affirm the world-picture on the basis of past and present experiences, whilst projecting meanings and values into the future in order to become increasingly confident in both the goodness of the modern world and the individual's personal capacity to deal with the future. The world-picture situates and orientates individuals in their participation in the modern world in such a way as to define the modern conception of the rational individual to be a willing participant and affirmer of the modern world. Those that will not or cannot participate are taken to be irrational, backward, or redundant. The modern individual has an increasing stock of innovative technology at one's disposal, whilst the providers of these technologies represent the modern good-life as being defined in terms of access to new innovations. The ancient conception of prudence or practical wisdom (*phronesis*), the ability to know from experience what to do for the best in any given situation or circumstance, has been transformed to how to choose between and acquire technologies to enhance one's life. The possession of the technical skill and know-how (*techne*) has become the highest virtue. Science and technology are inextricably bound together into a societal endeavour that permits technology to disclose reality to modern humans, while it is the very instrumentality of these disclosures – both to further exploration and practical problems – that provides the sense of being in touch with reality. The world-picture provides references through which we can scientifically evaluate, order, and interpret our experiences by projecting a framework of meaning over the events in our lives, in such a way as to narrate connections between individual lives, particular events, and with the whole Universe. Thus our lives can be represented as connected with the unfolding of the greater whole and the interplay between cosmic forces. The world-picture allows the individual to be situated and orientated with a Universe that transcends the individual, whilst simultaneously connecting the individual's nature and agency with Nature. Human flourishing is represented as the technological liberation of the individual from circumstantial constraints that inhibit the emergence of his or her inborn potential. The world-picture allows technology to be represented as the provider of opportunities for the emergence of human emancipation, character, and flourishing. Ironically, the world-picture simultaneously liberates human beings from causal responsibility, whilst representing the enhancement of power and freedom through technology as being a moral good and obligation. The participation in the construction of the technological society is represented as a rational and moral act, and any resistance to this project is represented as irrational or immoral should it impede or pervert the innovation and dissemination of scientific knowledge and technology. Modern science is a value-laden, moral duty to participate to the best of our ability to the innovation of the universal means to remove all of the world's evils and ills. The innovation of new means brings with it the moral obligation to use them, make them universally available, and to improve them. Henceforth, we all have the duty to acquire and use the best technologies at our disposal and, to the full extent that we are able to, to aid in the project of innovating better means and more technical knowledge. Ironically, the scientific world-picture represents the world as something at our disposal, but our affirmation and participation in it puts us at its disposal.

Technoscience and the technological society

The term "technoscience" is frequently used to highlight the transformation in the nature of science that has occurred since the end of the Second World War.[32] Modern scientific research increasingly requires large teams of researchers, enormous amounts of money, and the aid of industry. Scientific research is increasingly technical work and "pure science" is "applied technology" in the sense that science not only requires technology to test and validate its theories and hypotheses, but science is itself a technology that is applied to problems. The term "technoscience" is used to highlight the intimate connection between science and technology. Thus, it highlights the extent that the content and direction of natural science is bound up with the wider character of the society from which it is emergent and within which it is situated. The idea of technology as "applied science" is one that is based on a total neglect to pay any empirical attention to how the "pure sciences" are actually researched and practiced. In many respects, technosciences such as physics, chemistry, and biology are applied technologies used to solve economic, military, and political problems and to justify social orders, prejudices, and inequalities. The modern conceptions of rationality, objectivity, and knowledge are politically and economically driven and shaped instruments of prediction, manipulation, and control. These have shaped technical decisions and the acceptance and interpretation of scientific results. The scientific quest is no longer a matter of educated individuals developing the theoretical understanding of Nature and mapping the Universe but, instead, is a large-scale collaboration between institutions, commerce, governments, and military agencies, which is directed to provide techniques for the ongoing process of ordering, controlling, and exploiting the world. The evolution of any technoscience is shaped by social, political, and economic demands and, in turn, technosciences shape political, economic, and social possibilities. By bringing new things into the world, such as hydrogen bombs, antibiotics, contraceptives, radio, motorcars, and genetic engineering, the world is irreversibly transformed. New political, economic, and social possibilities arise when new technologies become possible. These possibilities, when realised, shape the directions of the technosciences and society. The evolution of science, technology, and modern society are intimately bound together to the extent that the modern world has become a grand experiment in human productive and manipulative possibilities. However, as I argued in *On the Metaphysics of Experimental Physics*, experimental physics has been a technoscience, bound up with commercial, civic, and military ambitions, as well as understanding Nature in technological terms, since its origins in the sixteenth century. In this book, I argue that this is also true of chemistry, medicine, and, from the nineteenth century onwards, biology as well. Modern science has always been intimately bound together with technology. The period since the end of the Second World War is simply an intensification of this intimate relation between science, technology, and the rest of society, rather than a radical transformation. Technoscience transforms the world and therefore must be analysed in terms of its broad aspects, including its social, political, economic, and ecological aspects, and not simply in narrow technical terms. We need to understand how it was possible, as well as understanding the real changes in social structures and organisation that occur as a result of technoscience, by examining the way that it conditions, constitutes, and mediates modern life. Hence, I agree with Donna Haraway when she defined technoscience as a form of life, a practice, a culture, a generative matrix formed by "dense nodes of human and nonhuman actors that are brought into alliance by the material, social, and semiotic technologies through which what will count as Nature and as matters of fact gets constituted for – and by – millions of people".[33] Technoscience emerged from the Enlightenment expectations of happiness and progress to arise through basing society upon the ongoing development of the practical and quantified sciences. It constructs Nature, mirroring the economic and political relations implicit to the social order, and represents Nature in scientific terms in order to make those relations normative and universal.[34] The modern world is constructed through technoscience by creating mixtures of active transgressions, mutations, hybrids, or cyborgs, rather than merely corrupting Nature by representing it in terms of cultural phenomena, which oppress and dominate the world, dissolving distinctions between natural and social, subjects and objects, natural and artificial, physical and abstract, real and simulated.[35]

Within modern society, experimentation and explanation are represented as defining and natural human activities. Technoscience provides the individual with resources for the construction of an autobiographical frame of meaning in which the individual can represent oneself in terms of the acquisition of technological powers, expertise, and agency, and connect one's life with the rest of society and the Universe as a whole. The individual is represented as a composite of juxtaposed functions and operational contexts, and others evaluate the success and failure of an individual on the basis of their function and instrumental value for the acquisition of further technological power, expertise, and agency. The individual is represented as a personification of agency and, in so far as uniqueness is acknowledged, one is only a unique mixture and ensemble of chemicals, combined through the operation of natural mechanisms. Thus individuals are represented as a product of their genes and circumstances, each a counterfactual of the same elements, laws, and mechanisms, for which any absence of full affirmation and participation from any individual in the moral obligations of progress is represented as a social or genetic disfunction that explains how such an individual anomaly is possible and suggests possible methods to remedy it. Once the social order is metaphysically conceived to be a mirror and consequence of the natural order, then the reluctance or refusal of any individual to fully participate within that society to optimal levels of efficiency (scientifically defined in accordance with the species-specific norms of human beings) must be represented as the natural consequence of natural mechanisms that impede that individual from achieving norms of human potentiality. It is imperative that these mechanisms must be discovered and suppressed for the benefit of the individual and society because there is the moral obligation that such an individual must be cured or prevented from happening again. Otherwise the individual will be inefficient and disadvantaged as an agent. This moral obligation to work towards technological efficiency and innovation is itself represented as the natural consequence of natural mechanisms that has, through the mechanism of natural selection, emerged as a species-specific norm because of its advantage for human survival. Of course, as the American pragmatists (such as Charles Saunders Peirce, William James, and John Dewey) pointed out, the everyday function of scientific interpretation is to offer a framework in which human beings can orientate themselves within their lifeworld. In our technological society, scientific reasoning and "common sense" must closely resemble each other to the extent that the former is widely considered to be a formal version of the latter. Moreover, scientists are situated people within society and the contents and directions of their work are shaped by political, economic, and personal considerations. Many kinds of non-scientific judgement influence the decisions about directions of research and the publication of results. Scientists, as well as the funders and publishers of science, *qua* people, base these judgements on what they consider their obligations to be, what they take as being human goods and values, and also what they consider to be the pressing problems and evils of the world. These considerations do not come from an isolated individual, à la Robinson Crusoe (as Marx quipped), but rather are the considerations of socially and historically situated individuals.[36] The lifeworld of individual scientists shapes and directs scientific research and hence science itself, which in turn is used by individuals to interpret their lifeworld, on the basis of either scientific reasoning or common sense, and orientate their choices and actions within it. The situation is one of feedback, refinement, and mutual reinforcement. The scientific world-picture and its representations of Nature are used to interpret the lifeworld of individuals, which in turn refines and reinforces the scientific world-picture. Nature is represented not only as being at our disposal, but also as being a threat to our lifeworld that requires being transformed into something that should be at our disposal more than it already is. While Nature is understood in technological terms, it is also represented as being hostile and aloof, thus reinforcing the need to transform and reduce Nature into something that is at the mercy of technology and best understood in technological terms.

In *The Technological Society*, Jacque Ellul described modern society as being premised upon a societal gamble in the rationality of the construction of society in the form of a technological society.[37] This gamble was that science and technology offered the means by which human beings could construct an artificial world that supposedly would be better than the natural one. This idea stemmed from the sixteenth and seventeenth centuries' promise of a disease- and poverty-free world, full of splendid bounty, based upon the knowledge and powers derived from mathematical and natural science. Since the nineteenth-century Industrial Revolution, the modern conceptions of rationality, power, and progress have been dominated by an adherence to efficiency, technique, and technological innovation. This dominance has extended to every human activity including agriculture, labour, medicine, information, politics, warfare, and education. Ellul's 1960s' critique of the modern technological society seems even more relevant today with the innovation of biotechnology, genetic modification, and cloning. This latest manifestation of the societal gamble is that an artificial controlled and manipulated animal and plant genetics and biotechnology promises to provide better agriculture and rid the world of poverty, hunger, and disease. It is even claimed that genetically modified human beings would be better than natural ones. Ethical considerations are often treated as if they were merely based upon emotional, fearful, ignorant, or subjective reactions to new technologies and new knowledge. Moral objections to the implementation of these new powers in the real world are often treated as irrational and unscientific resistances to the inevitable march of progress and evolution. Ellul spelled out what he called the essentials of a sociological study of the problem of technology by examining how "Technique", the sum of all techniques, all means to unquestioned ends, impacts upon society. According to Ellul, Technique is the "new milieu" of modern society, replacing the natural world, and all social phenomena today are evaluated according to their instrumentality; all the ideals and values of modern society have been fundamentally transformed by technology, which supersedes political ideology and pervades all areas of social life.[38] The technological society develops autonomously. Ellul argued that, as a technical civilisation, the West is entirely constructed in terms of techniques to such an extent that only that which is technical is considered to be part of civilisation. Everything must serve a technical end and anything non-technical is either excluded as "inefficient", "subjective", or reduced to a technical form. Every intervention of technique is, in effect, a reduction of facts, forces, phenomena, means, and instruments to the schema of the technological imperative. The human agent becomes transformed into an agent that is defined in terms of his or her performance and function, as an integrated and articulated component, in an ensemble of functioning agents. Technique sets upon and organises human agency. The project of "the technical man" is a perpetual search for the "one best way" to achieve any designated objective, and the perpetually expanding and irreversible role of technique is extended to all domains of life. The choice of method/technique is made in reference to the satisfactory stabilisation of measurements, calculations, and productive practices, in relation to an intelligible causal account. Such a choice cannot be divorced from the cultural background against which it is made and emergent from. It is a matter of paradigmatic socio-technical consensus about what is "the most efficient" and "the one best way". Until it is replaced by another technique, every technique achieves a technical autonomy in practice because technical practitioners, also under the sway of the technical imperative, are obliged to use it. Its results are indisputable until a "better" technique takes its place. Once a technique has become established as "the most efficient", then it is no longer an object for technical deliberation. The technical practitioner is committed, whilst under the sway of the technical imperative, to perform his or her operations in "the most efficient" manner which, of course, requires using "the most efficient" technique. Ellul questioned the meaning of the dominance of technique for the human present and future, and narrated the tragedy of a civilisation increasingly dominated by technique. He placed an emphasis upon the erosion of moral values brought about by technicism, an examination of the role of technique in modern society, and a historical disclosure of the forces that have shaped the development of technical civilisation. Modern society is a progressively technical one committed to the quest for continually improved means to carelessly examined ends. What was once prized as a good in itself, for its own sake, is transformed into something that is only of instrumental value, for the achievement of something else. Whilst technology transforms ends into means, human beings are compelled to adapt to a technical substratum of human existence that has become so overwhelmingly immense that we are unable to cope with it as a means and, consequently, treat it as an end.[39]

In this book, I am using Ellul's broad definition of technology, rather than limiting it to being defined as the technical use of devices or machines. Technology is any socially organised collection of knowledge, skills, tools, materials, devices, machines, and practices of intervening in the world in order to change the world. Technology is not just hardware. Hardware without technology is junk. Technology is a creative and social endeavour which provides the means to achieve otherwise unreachable goals. The values, aspirations, and ideologies of society are aesthetically built into the design of its architecture and technological infrastructure.[40] All technologies embody human values and purposes in their structure and design, and reproduce and disseminate those values and purposes within society. The messy and complicated employment of technologies is a consequence of the diverse and often incompatible political, economic, and aesthetic values, aspirations, and choices involved in the technological construction of our society. The traditional view that technology is a tool and science is a method is one that is based upon an idealisation, an abstraction, of the real situation. Human beings do not simply use technologies, but, in modern societies, to put it much nearer the mark, human beings have become technological beings. Technology brings substance to our desires, priorities, and necessities, which then become objective parts of the world. It does not simply change the material structure and arrangement of the world; it also changes what we do, and, therefore, at least in part, changes who we are. New technologies profoundly transform human agency and character. Technological innovation creates new ontological opportunities and demands for people to achieve new goals and to perform tasks in different ways. It has transformed the human condition and our understanding of what humanity is. Technology is not just an instrument, a means to an end, but is also a mode of expression, a self-defining way of being in the world. Technology is an evolutionary trajectory that is defined as its operations adjust to the specifics of its environment, which include social, political, and economic dimensions, as well the technical problems posed by its growth and its competition with other technologies.[41] It is the process of using and modifying technology to solve particular problems – many of which are unforeseen by the designers of technology – which gives technology its shape and content. The purpose and function of any technology cannot be separated from its operation in society. Technical rationality is a socially bounded and contingent form of rationality directed in accordance with contextual applications of the concept of "efficiency" to technical judgements. Hence, every technical decision about efficiency is dependent upon the social context within which that decision was made. All technical decisions about what is the most efficient technologies available are generated, structured, and reproduced as a result of a historical struggle and social process of establishing and disseminating societal goals, meanings, values, and choices, by suppressing alternatives and resistances.[42] The horizon of possibilities and the ways to reach it are reiteratively shaped by the social, political, and economic choices and problems, for which technology has promised to solve. New political, economic, and social possibilities arise because new modes of social organisation become possible. These possibilities, when realised, shape the directions of technology. Technologies are non-linear and complex, emergent through interconnected socio-technical feedback relations emerging from an evolving socio-technological background, and are no more neutral than the contexts in which they arise because the uses to which any technologies are put to are the content of their evolution. When modern knowledge is founded upon material practices, scientific rationality is transformed into bounded technical rationality, which is itself based on a gamble in the goodness of any particular vision of how society should develop. Any inequality in the dissemination of technology and scientific knowledge excludes some people from participating in decision-making processes and the establishment, or imposes the values and purposes of a social hierarchy and class structure into the implementation and development of the technological basis of our existence. Thus technical rationality entails ethical and existential judgements. How do we understand ourselves as human beings? To what extent are we defined by what we do? What role does technology have in our understanding of who we are? How does technology structure our actions and participation in society? The existential importance of questioning technology becomes clear once we address the extent that technology transforms human agency, which in turn transforms how we understand ourselves as agents in the world, and technical judgements are themselves socially bounded and evolving judgements. It is for this reason that the traditional view of technology as being simply a means to an end is, as Heidegger pointed out, the worst view possible because it immediately puts us in an unthinking relationship with technology.[43] We must make special efforts to critically raise our awareness of the extent that our ethical and philosophical character, as Heidegger put it, is already "delivered over to mass society [and] can attain reliable constancy only by gathering and ordering all their plans and activities in a way that corresponds to technology".[44] The exchange of propaganda has taken the place of a debate about ideas and it is widely taken for granted that the current development of society is the only way that society could develop.[45]

In his famous book *Technics and Civilization* (first published in 1934), Lewis Mumford criticised the traditional academic neglect of technology.[46] He argued that it was necessary to understand technology in order to understand modern society and our lives. Mumford's views about technology were deeply influenced by Marx, but he was critical of Marx's assumption that technology (as a means of production) was comprised of neutral technical forces that evolved automatically and determined the character of all modern institutions, whereas Mumford considered technology to be a contingent embodiment of human choices. He argued that the relationship between social institutions and technology was reciprocal and many-sided.

Technics and civilization as a whole are the result of human choices and aptitudes and strivings, deliberate as well as unconscious, often irrational when apparently they are most objective and scientific: but even when they are uncontrollable they are not external. Choice manifests itself in society in small increments and moment-to-moment decisions as well as in long dramatic struggles; and he who does not see choice in the development of the machine merely betrays his incapacity to observe cumulative effects until they are bunched together so closely that they seem completely external and impersonal. No matter how completely technics relies upon the objective procedures of the sciences, it does not form an independent system, like the universe: it exists as an element in human culture and it promises well or ill as the social groups that exploit it promise well or ill.[47]

Mumford called for an organic transformation of technology in order to make it more effective, harmonious with our environment, and place it at the service of life, rather than merely at the disposal of the opportunistic schemes of capitalism. The intimate connections between aesthetic, architectural, ecological, social, and economic relations should be central to the development and integration of technology within our environment and lives. Of course, the coherent and integrated process of organically developing the technological basis of our society will be slower than the headlong rush to innovate and implement everything that merely offers the possibility of further profits and immediate practical advantage, but, on the other hand, a co-ordinated social effort to relate and integrate technological developments within society, in accordance with needs of the wider community, will produce a much more stable, balanced, and harmonious technological basis for society. In his later work, *Pentagon of Power*, Mumford was highly critical of the centralised military–industrial complex, operated in accordance with the interests of a social elite and its technical servants, and how it had produced a "mega-machine" that ran roughshod over all aspects of society and was perpetually threatening to destroy all life on Earth.[48] He argued that the history of modern society is the history of the construction of a technological utopia that has become realised as a dystopia. While Mumford agreed with Marx's view that each period of technological innovation and production had its own historic mission and that the value, meaning, and purpose of machines are given by their social use and implementation, he considered Marx as being overly deterministic about the autonomy of technology. According to Mumford, we are responsible for how we choose to develop the technological basis of our society. However, as I shall argue in this book, even though I agree with Mumford that our technological complex and its trajectories are dependent upon our choices, we need to address the extent that, once embedded in society, those technological complexes gain autonomy and require sustained choices in order to change or redirect them. Of course, I completely agree that it is important to be aware that bounded technical rational judgements are human choices, but it is also important to analyse how technology can constrain, resist, and even coerce these choices.

My argument in this book is that the possibility of developing a genuinely democratic society is dependent on our capacity to decide how to develop and implement science and technology. If we hope to be successful in this endeavour, then we must critically examine the pernicious and antidemocratic aspects of our technological infrastructure and the institutional processes through which that infrastructure is developed. This involves a deep critical examination of the political and bureaucratic processes through which technical decisions are made about which technologies we should implement and develop in our communities. This involves a close examination of the vision of society and the human good life that these decisions entail, and, where such a vision is lacking, we must question the rationality of those decisions and our complicity in them. Of course, there already is a long tradition of critical reflection on the pernicious and antidemocratic aspects of modern technology. After the end of the First World War, Werner Sombart argued that task of technology is to liberate human beings from the limitations of their natural, organic state, to transcend these natural limitations, and replace the natural world with an artificial one.[49] However, as a critic of Marxism, he maintained that cultural ideas and values defined technology, and thus he denied that productive technologies defined economic and cultural life. He asserted that technology and culture are distinct and that it is the cultural spirit of a society that directs and limits its technological development and its economy. He was critical of the way that we have overly focussed on the development of the technological infrastructure of society, while neglecting the development of its cultural spirit. Oswald Spengler was another early critic of modern technology as being a pernicious and barbaric development of Western Civilisation.[50] He predicted that the cultural obsession with power and the mechanisation of the world will result in widespread plunder and rape of the natural world, extensive deforestation, mass extinctions, the devastation of landscapes and communities, and would result in a global ecological crisis. Such critical thinkers began a philosophical movement directed to the critical evaluation of modern technology and the resistance to limiting our decisions to technical rationale. Carl Mitcham termed this tradition as the "humanist tradition".[51] As well as Mumford, Heidegger, and Ellul, Hannah Arendt, José Ortega Y Gasset, and Hans Jonas were all highly influential members of this tradition.[52] This tradition has mounted a sustained philosophical critique of the dominance of instrumental rationality and positivism within modern thought and politics. They were concerned with the way that modern thought has become obsessed with finding the most efficient means to achieving unconsidered ends. This way of thinking – if it can be really called that – has become so pervasive that all other ways of thinking are either completely suppressed or dismissed as subjectively irrational. In many respects, Albert Borgmann is the contemporary spokesman for this tradition. In *Technology and the Character of Contemporary Life*, he argued that the Enlightenment was in essence the promise of a technological utopia in which human beings would achieve happiness, liberty, and prosperity.[53] It is this promise that conditioned and constituted the origin and development of modern technology. Borgmann asks whether modern technology actually fulfils this promise or whether it erodes our traditional values and prevents us from developing a thinking relationship with our lives.

I wish to situate this book as a continuation of this tradition by developing the philosophical critique of modern technology and science, in order to promote the awareness of the complicity that we have in the way that our lives have become dominated by the technosciences that were supposed to liberate them. The argument of this book is that we need to examine technoscience as a societal enterprise directed to transform the human condition, liberating us from the limits of our organic state of animality, and construct an ordered and stable technological society in place of the capricious natural world. Unless we critically examine the visions of the human good life that are implicit to this project, we will not be able to develop the rational and pluralistic basis of society that is necessary for the emergence of a genuine democracy. My argument is that a participatory, democratic, and community-based society is necessary for the optimisation of diversity, creativity, and experimentation in the construction of the technological society, which is crucial for the health of society and improves our ability to respond to changes in the world and the capriciousness of Nature. Hence I agree with Jürgen Habermas' argument that technical rationality should be subordinated to a rational form of communication and purposive action that embraces moral and political deliberation.[54] However, before we can do this, we need to critically examine the extent that mechanical realism, the societal gamble, and the scientific world-picture have become so entrenched within our modern society, that it has become increasingly difficult to rationally criticise the naturalness and power of technical rationality. Mechanical realism presupposes that technology is based on a twofold relation between natural laws and contingent social choices: only those contingent choices that successfully utilise natural mechanisms in accordance with natural laws can be integrated into technoscience. It has become increasingly difficult for us to develop a rational consciousness of the social contingency of technoscience, without degenerating into nihilistic relativism or postmodernism, but once we address the metaphysical and ideological foundation of the technosciences, then we can become aware of the extent that the epistemological truth of the technosciences is contingent upon our acceptance of the ontological goodness of the technological society. Thus, one of the ironies of modern society is that positivism and technical rationality are premised upon metaphysical precepts and ideological goals. Modern society is a technological experiment based on a metaphysical and ideological interpretation of the world. If we can accept this insight, then it is clear that technical rationality entails moral presuppositions about the kind of world within which we wish to live. Technological innovation substantively transforms our available moral choices and, as moral agents, we need to examine the substantive moral significance of technology upon how we live our lives. However, the technological imperative is a moral imperative to construct the best of all possible worlds and we should question the assumption that this will only be possible if technoscientific innovation is given absolute priority and free reign. This assumption is based on a societal faith that any social problems caused by any technology will be solved by further technological innovation and any attempt to limit it would limit the chances that human beings will be able to solve future problems. Our lack of foresight is actually used as an argument for affirming the act of blindly leaping into the unknown. However, if we wish society to be rationally developed, then it is the faith in this gamble that must be brought under our critical gaze. This is what I aim to do in this book. Whether one has an optimistic or pessimistic view of the progress of technology, one needs some vision of an ideal society against which to make this judgement. Even if we understand ourselves, including our cultures and technologies, as being based on Nature and part of it, we still have to recognise and analyse the pluralistic and incommensurable possibilities that emerge from our nature. The problem of how to choose the basis by which we can decide between them is not resolved by appeals to the ontological naturalness of these possibilities. Even if we accept the ontological naturalness of technology, this does not help in the ethical dilemmas and moral problem of whether and how we should use technology to transform the world and develop society. What we need to do is to bring the aesthetic and existential possibilities of technology and its alethic modalities of production and innovation into the domain of democratic deliberation. It is more than just a question of whether we can rationally understand and direct the processes by which we understand and interact within the world through science and technology, but involves widespread societal participation in the question and debate about what a rational understanding and direction of technology and science would be. We need to not only examine how our technological activities and their consequences cohere and are sustainable, but we also need to examine the ideals (or absence of ideals) under which the whole process is directed in order to question whether that process is governed through rational intelligence, or if it is a self-generating juggernaut within which human beings are merely components and resources. The crucial step in the development of contemporary debates is to propose, articulate, and critique visions of an ideal society. Otherwise, our theories and policies are without direction and purpose, and are nothing more than the irrational products of social and intellectual conformity.

2
The Metaphysics of Modern Science

In this chapter I shall generalise my argument presented in *On the Metaphysics of Experimental Physics* to discuss the operational metaphysics of modern experimental science. It is one of the assertions of positivism that metaphysics has nothing to do with science. However, this assertion is a myth. Experimental physics – the exemplar of positivistic science – is premised upon the operational metaphysics of mechanical realism. This metaphysics was required for the conceptual establishment of a methodology to explore Nature; it provided a unifying conception of "the physical" that underwrote the foundational principles and assumptions justifying the technological enterprise of the experimental sciences as natural sciences. Thus the natural world has become experienced in technological terms whilst technology has become conceived as being based upon natural principles. The epistemology of experimental science presupposes that only those artificial means that function according to natural law are capable of functioning at all in virtue of their utilisation of natural mechanisms. Supposedly, the only technologies that are possible are the ones that are constructed in accordance with natural law. Hence, once mechanical realism is presupposed, any possible machine is not radically different in kind from any possible natural entity. They are merely counterfactuals from the same natural laws. Henceforth, natural phenomena can be represented, modelled, and understood in terms of machine performances, and our understanding of natural laws can be "tested" by applying it to technological innovation. Technological innovation has not only made new observations and experiments possible, but it has also transformed our experience and conception of reality. Hence the positivistic assertion that the content of scientific knowledge is confined to empirical statements is an assertion that cannot be readily derived from an analysis of actual scientific methodologies, discourse, and practices. Traditional philosophers of science try to maintain a clear distinction between "pure" and "applied" science. They assert that theories precede and anticipate experiment in the form of hypotheses, conjectures, or predictions, and that experiments are simply the means to test theories. They assume that science provides us with a rational understanding of Nature, embodied in technology as "applied science". This view presupposes that rational thought and logic transcend the material world and that the primary relationships between the human mind and the world are those of cognition, manipulation, and control. It also presumes that technology enhances and extends the powers of the human mind and senses without changing or directing either. The construction of theories is supposedly a purely intellectual affair for which the technology of experimentation does not have any constitutive scientific role. The experimental apparatus and methodology are represented as being ontologically and epistemologically neutral. Technology is ignored as largely irrelevant for the epistemology of science, as something that, at most, is a matter for applied ethics. However, as many critics of the traditional philosophers have taken considerable pains to show, this is evidently not the case.[1] The very character of science is transformed by the demands of society. Scientists are involved in an industry of economic exchanges within the society within which science is emergent, producing values and powers for the wider world of commercial, military, and political ambitions. Scientists may well maintain that they make and use their experiments and theories purely for the purpose of representing reality, but in the wider world they are used as weapons, tools, products, and displays of national or civic prowess. Social and cultural values determine the choices between which scientific projects and directions are funded, either from governmental or private sources, and the content of what is considered to be "rational" and "progressive" in the acceptance and justification of scientific work is determined through consensus rather than logic.

Empirical science is supposedly based on the facts of experience – upon scientific observations. The corroboration of theories occurs through testing its hypotheses against the facts of experience. Consequently, a theory must predict detectable effects that can be tested if it is to be considered as a scientific theory. However, in all but the most trivial cases, observation in science requires technological mediation, using instruments, detectors, sensors, probes, and so on. Often the outputs of instruments are mediated using computers and models before they can be presented in an intelligible form on a screen or graph. These observations are indirect and require theoretical representations and models of a causal chain of events and mechanisms between the observer and the observed. It is in this sense that observations can be considered to be theory-laden. Moreover, for the contemporary sciences (especially when investigating phenomena that either do not occur naturally – such as superfluidity – or involve microscopic features and structures – such as genes) techniques of intervention are required in order to make the observations in the first instance. Theoretical representations and models of the experimental apparatus or measuring instrument are required to make observations. In this sense also, observations are theory-laden and mediated by technology. Traditional philosophers of science have neglected to examine the way that artificial devices have been used to understand the natural world. This neglect has led to a false dichotomy, which has largely restricted the debate to being an argument between anti-realist and realist interpretations of physics. Consequently, the knowledge produced through experimental physics is supposedly *either* of human origin (science is situated in a context of justification) *or* it discloses objective structures of the natural world (science is situated in a context of discovery). For the realist, it is self-evident that Nature must be the non-human participant that resists, empowers, and inspires human intentions and theories. Thus, the realist criticism of anti-realism is that it fails to address that there must be both human and natural aspects at play in experimentation in order to explain the failures and successes of science. However, once we examine the substantive transformation that technology has on human agency, then the traditional assumption that technology is simply a "man made" means to human given ends is problematical. Technology transcends and structures individual human action, directing and empowering human beings as causal agents within a technological framework, within which scientific activity is situated and shaped. The meaning of any human intervention can only be theoretically understood in terms of its function within the structure of the technological framework. It is made meaningful in terms of an abstract representation of its functionality mapped into an abstraction of a series of procedures and their associated consequences, understood in terms of expectations and intentions. Thus any description of function is an evaluative one that is made in terms of an expected norm or postulated purpose. Hence, material and representational practices cannot be separated from the other within the technological framework, and, therefore, scientists cannot separate their experience of what we are observing from their theoretical understanding of how the experiment works. It is for this reason that the theoretical understanding of the properties of natural entities, represented in terms of their functionality within the operation of the experiment, is one that is a serious problem for the traditional philosopher who presupposes that the structural properties of natural entities cannot be understood in terms of functionality, given that function is always extrinsically imposed on an object by an observer. Once we recognise this problem, then it makes the traditional presumption that simple experiments provide facts for theories completely untenable. The distinction between the functions and physical properties discovered through experimentation is only possible if the technological context is ignored and properties are abstracted from the circumstances of their identification. Properties of hardness, durability, temperature, instability, and so on are all relational properties that are identified in relation to particular technological interventions, made in terms of mutually transforming and ordering techniques and theoretical representations, tested in terms of the success or failure that results from attempts to integrate them into the already established technological framework. This technological framework selects and orders intellectual and material practices in accordance with procedures and expectations that precede the work of individual scientists. Accordingly, individual scientists do not control the outcome of their experiments and the results of experimentation do not always conform to theoretical expectations. The task of integrating novel scientific work within this technological framework is one of innovating machine prototypes and/or techniques, according to the already established demands and expectations of society, while representing the ontological possibility of the alethic modalities of human agency in terms of natural mechanisms, operational within a context of discovery. At all stages of experimentation, the objects of scientific practice and discourse are machine performances, given in terms of responses, measurements, resistances, and limits, which are represented as being the products of the rational realisation of natural mechanisms in material practices. The knowledge produced from experimentation is the theoretical interpretation of the performativity of machines, therefore, experimental sciences can be understood as the social products of innovative, artificial processes, within which discoveries are machine performances, which are then used to justify theoretical interpretations of them. Furthermore, this explains how experimental sciences achieve considerable predictive success and innovative progress, and how the representations used in scientific work can be rhetorically placed in correspondence to underlying natural structures or mechanisms.

Many historians of science have argued that the practical arts led the sixteenth-century "scientific revolution" and that modern science is "applied technology" to a lesser or greater degree.[2] However, as I argued in *On the Metaphysics of Experimental Physics*, the characterisation of the natural experimental sciences, such as physics, as "applied technology" only reverses the problem of how the "application" occurred.[3] We need to analyse sciences, such as physics, at a deeper level than merely pointing out that the use and development of technology has been central to the experimental natural sciences since their origin in the sixteenth century. We need to historically trace back the origin of these sciences in order to understand how these technosciences could be represented as natural sciences. Mechanical realism underlies the whole epistemological legitimacy of the technological disclosure of natural mechanisms in the experimental sciences. This metaphysics was established during the fifteenth and sixteenth centuries and made the "scientific revolution", experimental physics, and modern technology conceptually possible. Mechanical realism allowed the mathematical description of the motions of the six simple machines (the wedge, the lever, the balance, the inclined plane, the screw, and the wheel) to be represented as descriptions of the fundamental natural motions. The Renaissance developments of the Medieval mathematical science of mechanics, underwritten by mechanical realism, allowed experimental physics to be metaphysically operational as a technological mode of disclosure of natural mechanisms, and for technology to be a consequence of the utilisation of natural mechanisms in material practices. This provided the sixteenth and seventeenth centuries with both a methodological and an ontological foundation for the mechanical and experimental natural philosophies of Galileo, Descartes, Bacon, Gassendi, Newton, Boyle, and Hobbes, *et al*. This methodological and ontological foundation was central to the methodology, intelligibility, and subsequent researches of experimental physicists. It made scientific naturalism possible. The discipline of physics grounds itself not in a phenomenology of our experiences of the natural world but in the abstract conceptions of "the physical world" that its practices make available for technological innovation. To the extent that these conceptions are robust, reliable, and valid, physics provides itself with firm foundations and can become in turn the foundation of further productive activity, which leads to the representation of technological innovation as being a process of discovering ontological depth. Mechanical realism made this self-reference possible and epistemologically grounded experimental science in machine performativity. It also explains how physics was able to achieve justification on the basis of its practical success in solving technological problems. The precepts of mechanical realism permit the fundamental relationship between human beings and Nature to being represented as technological – hence bounded technical rationality and interventional material practices become the fundamental mediations between natural beings and human beings. The modern conception of the individual became possible and the world could become represented in terms of a theory of the real in terms of "natural mechanisms" related in accordance with "natural laws". The world is represented in such a way that it becomes objectively grasped in terms of a world-picture, and the technological society is represented as the rational consequence of the application of scientific knowledge to human needs. The traditional realist and positivistic philosophies of science already presuppose the precepts of mechanical realism within their epistemologies and ontologies. This presupposition has become so deeply entrenched in our modern world-view that any anti-realist philosophy, which criticises realist or positivist philosophies of science, is open to the criticism that it is simply based on mysticism, irrationalism, or relativism. The metaphysical precepts of mechanical realism have become so deeply entrenched in the modern world-view that they are not even considered to be metaphysical precepts – they are considered to be self-evident and based on practical common sense. In this chapter I shall describe how the connection between the artificial and the natural was based on the metaphysics of mechanical realism.

Mechanical realism and the scientific revolution

Aristotle, following Plato, defined *techne* (plural: *technai*) as a kind of theoretical knowledge of the explanatory, generalised, abstract, formal, and communicable first principles (or intelligible causes) about how to perform arts and crafts.[4] *Techne* provided "a true course of reasoning" about how to make particular things in a specific manner that was inextricably bound up with an intellectual grasp of first causes that provided the kind of knowledge possessed by an expert (*technite*) in any one of the specialised crafts, which guided stable dispositions to make particular things or bring about a state of affairs in a specific manner.[5] *Techne* was the general theoretical knowledge of the changeable and temporal, as distinct from *episteme*, which was reserved from the general theoretical knowledge of the unchangeable and eternal. As a general knowledge, *techne* was used to explain the particulars of experience, during activities of making, but it was taken to be distinct from them. While habitual practices (*praxis*) could be learnt from mimicry and trial and error, allowing the development of tacit (non-verbal) skills, it was only when the person acquired a complete explanatory, rational account of *praxis* that they could be said to possess *techne* and be considered as a *technite*. Within Aristotle's fourfold causality of formal, material, final, and efficient causes, it was the *technite* who took on the role of efficient cause. *Techne* guided the hands to perform definite motions that moved the tools and imposed form (*eidos*) into matter (*hyle*). *Hyle* referred to an unknowable, formlessness potential to resist the imposition of *eidos*.[6] The *technite* had to be responsive to the way that *hyle* responded to the imposition of *eidos* and the extent to which the *technite* could impose it was not entirely within the control of the *technite*. It is only to the extent that the imposition can be grasped by "the rational part of the soul", as *eidos*, that it can be known and a part of *techne*.[7] The *technite* must attend to *hyle* and be responsive to the way the particularities of the particular resists the imposition of the generality of *eidos* from having complete sway, and the *technite* needs to use perception (especially touch) to guide the activities of making.[8] Thus, for Aristotle, theory is an incomplete guide for action, and productive activity (*poiesis*) guided by *techne* was straddled on a continuum between particularity of practice and the generality of theory.[9] The role of *techne* was to facilitate the tracing back of a product to its causes, as well as recognising the appropriate forms, tools, and materials to be used to govern and guide *poiesis*.[10] *Poiesis* was taken to bring about and terminate in a product, outcome, or end (*telos*) that was separate from *techne* and *praxis*. A pot is brought forth, in accordance with the clay, the actions of the potter, the form of the pot, and the purpose of the pot, whereas a tree is brought forth in accordance with an internal principle of change (*phusis*). Even though Aristotle used his conception of *techne* as his primary analogy in his elucidation of *phusis*, he maintained the autonomy of *phusis* and considered *poiesis* to be distinct from *phusis*.[11] Aristotle made a distinction between things that find their origin in the maker (*poieta*) and things that find their origin in themselves (*phusika*). When *telos* was introduced through the activity of a *technite*, the source of change was separate from the thing in which the change happens; *phusika* had their source of change immanent as an internal principle, hence the *telos* of *phusika* were directed according to their essential nature, within their natural place. Consequently, Aristotle based physics upon a phenomenological account of natural phenomena in their natural environment, rather than experimentation under artificial conditions.

Experimentation would teach us nothing about the nature of natural substances and beings because any attempt to isolate a being from its natural environment would inhibit, obstruct, or change its behaviour.[12] For Aristotle, art imitates Nature; the products of the craftsman were able to capture the semblance, but not the essence of the natural exemplar that is copied.[13] The Aristotelian interpretation of mechanics was that it was an art that was working against Nature by contravening the natural tendencies of materials by subjecting them to artificial, violent motions.[14] Prior to the "scientific revolution", *scientia* had contemplation as its goal and the practical arts had making and operating as their goal. Aristotelian *scientia* was based on the contemplation of ordinary experience of natural phenomena and thereby the contrived experiences of experiment could not disclose anything about Nature.

Francis Bacon turned this view on its head when he asserted that artificial products and natural beings differ not in essence, but only in their efficient cause.[15] Bacon's experimental philosophy was based on the idea that it was possible to discover natural causes by artificially reproducing the phenomenon in question. For Aristotelians, artifice was unable to equal or approach the essence and subtlety of natural processes, whereas for Bacon, knowledge of natural phenomena could best be acquired by replicating them. Thus, he transformed Aristotelian *scientia* by using its concepts, definitions, and categories in order to develop a self-conscious "maker's knowledge" of Nature. Rising upon the ascendancy of the mechanical arts, Bacon was able to develop a concept of *natura vexata* (Nature as constrained, moulded, or interrogated by art and experiment) as opposed to *natura in cursu* (Nature doing its own thing in an unrestrained state), and thereby he overthrew the sharp Aristotelian distinction between *poiesis* and *phusis*. Bacon proposed that natural philosophy should not rest content with observing a free and unconstrained Nature, but instead asserted that it should be a part of the practical arts, based on observations of Nature "constrained and harassed when it is forced from its own condition by art and human agency, and pressured and moulded".[16] He argued that the natural axioms induced from experience founded the mechanical arts, which he praised for providing a "variety of objects and splendid equipment", having "contributed to human civilisation", and being based on "axioms of nature" discovered by observation and subtle, patient, ordered movement of hands and tools. If directed according to utility, they were capable of growth and flourishing. He cited the clock as an example of "a subtle and precise thing that seems to imitate the celestial bodies in its wheels, and the heartbeat of animals in its constant, ordered motion; and yet it depends on just one or two axioms of nature".[17] He considered the mechanical arts to be praiseworthy as the source of civilisation and political advantage in general and the discovery of the art of printing, gunpowder, and the nautical compass in particular. Mechanical art was the noblest human pursuit and "right reason and sound religion would govern its use".[18] However, how was Bacon able to achieve this conceptual revolution upon the ascendancy of the mechanical arts? How was he able to assert that "maker's knowledge" was the knowledge of natural principles or laws? We need to examine the culture from which Bacon's natural philosophy emerged. In *On the Metaphysics of Experimental Physics*, I argued that the "scientific revolution" was a mechanists' revolution within which the experimental sciences grew out of the efforts to transform the status of mechanics from the banausic into the scientific through the mathematical rationalisation of machines.[19] The demand for a new world-view within which all phenomena could be connected and explained in terms of underlying mechanisms, described in accordance with geometrical representations, was based upon the successes of the fifteenth-century mechanical arts in satisfying commercial, military, and civic ambitions. It was the ability to abstract the machine performances of the six simple machines that empowered the confidence in the rational reduction of mathematics into forms that were capable of being implemented into technology, which preceded the metaphysical reification of the results of the experimental sciences into abstract natural laws. Bacon's natural philosophy was emergent from a cultural affirmation of the practical arts, within which the sixteenth-century artisans saw all knowledge as rooted in material practices.[20] Maker's knowledge became transformed into an informal, tacit *techne* through which the *episteme* of how Nature works could be disclosed through intervening into natural processes and constructing mechanical models of natural entities. Bacon's natural philosophy was based on mechanical realism. Once mechanical realism had been established, then *natura in cursu* could be revealed through *natura vexata*, and thereby experimental philosophy allowed the process of making and operating to be the goal of the contemplation of Nature, and therefore was a natural philosophy. Accordingly, for Bacon, manual art was simply human beings added to Nature and, given that human beings are natural, then knowledge of the manual arts was knowledge of natural principles. As Bacon put it: "All man can do to achieve results is to bring natural bodies together and take them apart; Nature does the rest internally."[21]

Mechanical realism first began to emerge in the philosophical writings of sixteenth-century Italian mathematicians, such as Nicolò Tartaglia,

Francesco Maurolico, Guido Baldo, Giulio Savorgan, and Bernardino Baldi. Even though these mathematicians maintained the Aristotelian classification of mechanics as an art, they argued that mechanics had to be based on natural principles – described in terms of geometrical proofs – in order to work in the real world. However, one of the most influential sources (and much neglected by scholars) for the emergence of mechanical realism is that of Guiseppe Moletti. Moletti was the professor of mathematics between 1577 and 1588 at the University at Padua.[22] Moletti followed the Aristotelian classification of mechanics as "contemplative philosophy" of mathematical principles of statics, dynamics, and kinematics. According to Moletti, the task of mechanics was to demonstrate the most efficient means of performing the maximum amount of work with the minimum of effort. For Moletti, mechanics was a science and not an art because the geometrical first principles of mechanics were necessary and eternal causes and truths, whereas the arts were contingent upon human desires and ends. He argued that the first principles of mechanics were natural means, that mechanics was to be found in all the works of Nature, and the first principles were "Natural Laws". Moletti formally reclassified mechanics to be a natural science based upon natural principles.[23] In 1592, Galileo succeeded Moletti as professor of mathematics at Padua. Galileo was trained as an artisan and an engineer, rather than a philosopher or mathematician, and maintained close ties with the mechanists, instrument makers, and craftsmen of Venice and Florence.[24] For Galileo, if one truly understood any phenomenon, then one should be able to construct a machine to reproduce that phenomenon.[25] Mechanical realism was implicit to his scientific method because for the natural philosopher to understand the true cause of natural phenomena he or she must be able to replicate or reproduce the natural phenomena by constructing an artificial device. Galileo went further than Moletti. Not only were the motions of simple mechanical devices based on natural laws, as Moletti had proposed, but they were also to be used to investigate natural laws. It was the geometrical treatments of simple mechanisms that were to be classified as natural laws and all natural movements were to be treated as the results of simple mechanisms at work. Galileo argued that the efficient causes of mechanics were the necessary causes and fundamental mechanisms of Nature, and the mechanical was to be taken literally as an embodiment of mathematics in the world.[26] In the metaphysics of Galileo, the transition between mathematical mechanics and the development of mathematical natural science required a set of precepts in order to appeal to a generalised principle of operation

in Nature in order to correlate the motion of bodies, and their properties, with measurements.[27] Galileo's mechanical realism was a necessary
precursor to the mechanical philosophies of the seventeenth century.
It was essential for the development of Galileo's new physics and was
to provide the methodological template for all subsequent physics by
projecting the geometrical abstraction of the balance over all natural
movement.[28] As a consequence of this reduction, all the natural laws
disclosed by all subsequent experimental physics, including the Laws
of Conservation (mass, charge, energy), Newton's Laws of Motion, the
First Law of Thermodynamics, and the applicability of differential equations to change in natural processes, are premised upon the explanatory
power of Galileo's metaphor of the balance as a fundamental mechanical
principle of Nature.[29]

Descartes' definition of a methodology for the new science was
profoundly influenced by Galileo's work and the same precepts of
mechanical realism underwrote his metaphysical foundation of natural
philosophy. Descartes' *Meditationes de prima philosophia* contains his
demonstrations of the metaphysical foundation of his methodology that
he proposed as a replacement of the Aristotelian natural philosophy in
order to free human beings from the bonds of medieval scholasticism.[30]
Descartes' natural philosophy reflected the passion for this new cultural
enterprise, which not only emphasised that mathematical deductions
based on clear and insightful intuition are the routes to knowledge,
but also held that method, in general, *mathesis universalis*, is necessary
for us to have truths at all. This method was to consist in the order
and arrangement upon that which "the sharp vision of the mind" is
to be directed if truth is to be discovered. It needed to be intuitively
self-evident, establishing in advance upon indubitable *a priori* principles
what constitutes being and from where, and how, the essence of being
is to be determined. His aim was to establish a mechanical philosophy
of Nature within which all natural phenomena could be explained in
terms of the extension and motion of innate matter in geometrical space.
Given that mathematical truths are clear and distinct, they must provide
truths of the physical world. Descartes used this reasoning to establish his characterisation of matter in terms of geometrical extension,
infinite divisibility, and primary and secondary qualities. These charac-
terisations constituted the fundamental elements of the physical world
within his natural philosophy. God was the first cause of motion and
always conserved an equal quantity of motion in the Universe according
to laws of inertia and impact. Descartes appealed to the perfection of
God in order to justify the possibility of *a priori* knowledge of these

laws of Nature, and, as a consequence, the same laws of Nature would govern any world created by God, and therefore in order to obtain knowledge of this particular world more than just the *a priori* laws of Nature are required. Physics was nothing more than the projection of mechanical science as the template for the geometrical description of all natural motion and change. Descartes considered the new physics to be mechanics and considered all natural phenomena to be machines.[31] The first principles of this science could be understood in terms of *a priori* axioms, leading to the identification of essences, and observation was only necessary to determine the contingent actuality of phenomena. For Descartes, knowledge of the laws of Nature was necessary but not sufficient to explain particular phenomena. Observation and experiment were needed to explain the phenomena of Nature because we needed to know which of all the possible phenomena are existent in this world and which of the several possible mechanisms, compatible with the same natural law, governed the production of the phenomena in question. Even though the laws of Nature were eternal and necessary, the actuality of the phenomenal world was contingent because how God implemented the laws of Nature was contingent.[32] The laws of Nature represent the possibilities of God's choices between possible phenomena and mechanisms when making the actual physical world in accordance with natural laws. In other words, in order to know which mechanisms God used to make the phenomenon in question, as well as which phenomena God had made, one needed to observe and experiment. Experiments and observations were not designed to discover the laws of Nature, but to select from a set of possibilities and to show how the general laws applied to particular phenomena. Experiments and observation were not sources of empirical data for the induction of natural laws, nor were they to test theoretical hypotheses, but instead were used to eliminate the possible explanatory mechanisms derived from *a priori* natural laws, and were constrained in terms of what could be made or manipulated mechanically either in practice or in thought. Descartes' scientific method was to produce mechanical analogies (or models) derived from first principles that would produce the same phenomena observed to exist in the world. Observations and experiments could then be used to eliminate deduced mechanical models from the potentially infinite set and provide criteria by which judgements regarding which mechanisms were the actual mechanisms involved in the production of the phenomenon in question. By using "empirical evidence" to eliminate deduced possibilities, except one, Descartes hoped that the demonstrative character of his natural philosophy would be secured, and he proposed that the task of experiment was to determine which mechanisms were at work in producing the natural phenomenon in question by artificially reproducing it. An understanding of the application of the mathematical science of mechanics to the physical world constituted the basis for an understanding of the productive capabilities of God. By representing God as an artificer who made the world as a machine, Descartes went further than Galileo.[33] Furthermore, once we understood the productive techniques of God, then we too could become more God-like in our capacity to change and produce things in the physical world, and "make ourselves the masters and possessors of Nature".[34] Descartes' natural philosophy intimately bound together the technical capacity to make with the discovery and subsequent implementation of natural mechanisms. Henceforth, the philosophical test of knowledge claims became a demonstration of productive success and any failure to implement *a priori* knowledge in material practices was a failure of the craftsman, rather than a failure of the theory.[35]

Newton used similar arguments regarding his prisms, arguing that if a craftsman failed to produce a prism that showed seven colours then the craftsman lacked the skills to make it properly.[36] Newton was very much influenced by the natural philosophies as proposed by Galileo and Descartes, as well as others, including Beekman, Cavendish, Charleton, Digby, Gassendi, Hobbes, and Mersenne.[37] All these natural philosophies were implicitly premised on the precepts of mechanical realism. Consequently, Newton was able to assert that mechanics should be used to investigate all "the forces of Nature", to deduce the motions of the planets, comets, the moon, and the sea, as part of empirical philosophy, rather than limit mechanics to the manual arts.[38] His "Rules of Reasoning" are based on the precepts of mechanical realism. This allowed him to assume that there is a "natural economy of causes", demanding that the investigation of the causes of natural things is limited to those causes we identify as necessary and sufficient to explain the appearance of natural phenomena, assuming the invariance and universality of any cause–effect sequence we discover from experiment, due to the isotropic and homogeneous character of natural law. Thus Newton was able to assert that "the qualities of bodies are only known to us through experimentation" empirically informed us about the bodies upon which no experiment had been performed. Following on from Galileo and Descartes, Newton considered natural philosophy to be founded on geometry and considered it to be that part of universal mechanics that accurately proposes and demonstrates the art of measuring magnitudes and motion, and, hence, the science of mechanics would be the science of any kind of motion resulting from any force whatsoever. For Newton, all causes were mechanical causes. Robert Boyle, who sought to explain cold, heat, magnetism, and all other natural phenomena in terms of mechanical principles, also assumed mechanical realism when he proposed that the route to understanding "the mechanical affections of matter" by artificially reproducing the phenomenon under investigation by building a mechanical device.[39] The truth-status of any knowledge obtained from performing experiments and artificially reproducing natural phenomena was itself provision of the value of that knowledge in the construction of future experiments and the innovation of new machines. Robert Hooke described his natural philosophy as the real, the mechanical, and experimental philosophy.[40] He considered true natural philosophy to be a process of using machines and instruments to make interventions into the natural order of things in order to explain natural phenomena in terms of fundamental mechanical interactions. Hooke frequently used machines to model and illustrate the fundamental and natural principles of mechanical motions. Newton, Boyle, and Hooke's natural philosophies all represented machines as having explanatory power about Nature, and they continued Bacon's dream of deriving the rest of the phenomena of Nature from the same kind of reasoning from mechanical principles.

The radical conceptual development of "the scientific revolution" was the emergence of novel symmetry: both natural phenomena and technological powers were to be explained and manipulated according to the same kind of principles or laws. The causes of art and Nature were taken to be categorically identical. This was simultaneously a naturalisation of mechanisms and a mechanisation of Nature. This was situated within the framework of mathematics and was an operational metaphysical synthesis of natural philosophy and the science of mechanics that conceptually allowed an ontological equivalence between natural entities and machines. This was a precondition for the seventeenth-century clockwork universe, as well as the subsequent refinements of this world-picture, by introducing practicality and naturalism to interpretations and representations that would have been nonsensical to previous natural philosophers. The scientific world-picture embodied the precepts of mechanical realism and made the experimental use of mechanical devices to ascertain the fundamental mechanisms of Nature epistemologically intelligible. Henceforth, the ontology disclosed by the new experimental science was a tripartite ontology of law, mechanism, and object. The performance of any machine is conceived as being governed by natural laws, and, consequently, the technological process of designing, building, and using machines in the artificial world of the laboratory provides a means to disclose the fundamental mechanisms and laws of the natural world outside the laboratory. Mechanical realism provided the necessary conceptual connection between the closed, controlled, and simplified system of the experiment and the open, complex, and messy natural world to allow machines to act as the objective connection between the two. Within the experimental sciences, objectivity is supplied by technology, allowing the machine performance to be the object of experience. It is because a machine performance is a repeatable conjunction of events, and each event is construed, in advance, as one of the projected six simple mechanical motions, then performing an experiment applies the mathematical projection of abstract machine motions over the phenomenon. Whether the results have some immediate practical use is irrelevant, what is important is that they are represented and understood in technological terms from the onset. Within the scientific world-picture, there is one single most efficient mechanism in operation between any particular cause and its effect(s), termed as "the natural mechanism", and it was the allotted task of the natural experimental philosophies to find it for any particular cause–effect sequence. The metaphysics of mechanical realism provided the link between the mathematical sciences and the practical sciences. Henceforth, the distinction between "pure" and "applied" science is merely the distinction between finding this "natural mechanism" and implementing it in practical activity, but, given that the test of the reality of any proposed mechanism is made by using it within technological innovation, experimental science involves both the discovery and application of "natural mechanisms" through ongoing technological innovation. It is internally both "pure" and "applied", regardless of whether its products are actually used for solving practical problems in the wider world. The myth of "pure science" is based on the fact that many scientists pursue science simply motivated by curiosity into the workings of the natural world. However, this hides the fact that Nature has been understood in technological terms since "the scientific revolution" and the extent that "pure science" is "applied technology". The use of abstract mathematical theories and quantified models of science, which have predictive success and practical application in the innovation and description of the performance of novel machines, represented in correspondence to natural laws, shows how experimental science and modern technology have been metaphysically bound together since their onset in "the scientific revolution". Not only are the connections between causes and effects in the natural world conceivable in terms of fundamental mechanisms that are governed by natural laws, but also the mathematical descriptions of the fundamental mechanisms that allow any machine to work are conceivable as mathematical descriptions of natural mechanisms. This metaphysics allowed technique to be understood as limited by the physical world and as an intervention into this world by mobilising "the forces of Nature". Techniques bind together machine performances and theoretical causal accounts, based on representations of how the machine works. Experimental science reproduces and explains change by connecting techniques, machine performances, and causal accounts. The rules and laws abstracted from experimentation take the form of *techne*, describing the first causal principles of change at work, represented as "natural mechanisms" at work during technological activity, allowing this knowledge to be represented as *episteme*, by using mathematics. The facts of experience are the constants and variables at work in the design, construction, operation, and interpretation of machine performances, disclosing the empirical regularities of the experiment by mapping out the interactions between human interventions and machine performances. The ontology of the experimental sciences is circumscribed by machine performances, and the process of discovery in experimental physics is one of innovation of novel machine performances and their associated cluster of techniques, representations, and causal accounts. However, mechanical realism is necessary if physics is to participate in discovery rather than merely mapping out the alethic modalities of the experiment. Natural laws are abstracted from the estimated possibilities, actualities, necessities, impossibilities, and contingencies of the reproduced and quantified interactions between human interventions and machine performances. They are the productive possibilities of the permutations of those interactions and, consequently, are inherently technological and metaphysical.

Of course, once the practical value of the new sciences had been rhetorically secured to the (albeit limited) predictive successes of astronomy, representing the whole Universe as a mechanical device, then all natural phenomena were assumed to be understandable as mechanisms. Even the human body became represented as a machine (within which the soul was seated, as a controller) and through the development of Renaissance dissection and representation techniques the body was disclosed as a mechanical device, which differed from all other natural phenomena only in complexity, rather than in kind or essence.[41] Life itself became increasingly represented as the complex organisation of inanimate matter, given its vital spark by God. Hence, by the nineteenth century, mechanistic accounts of life had increasingly replaced the catalogues of natural historians, botanists, and zoologists, and all living beings could be represented as an internally organised machine: an organism. Life became understood increasingly through technical interventions. Thus, even though his conclusions were controversial, there was acquiescence about Charles Darwin's use of an analogy between the artificial selection techniques of pigeon breeders and natural variation in wild finches, in order to develop his explanatory concept of natural selection. Towards the end of the nineteenth century, biology was transformed into the experimental and laboratory study of embryology, physiology, and animal behaviour.[42] Biology transformed from the descriptive study of entities in the natural world into the mechanistic explanation and possible artificial reproduction of life itself. Whilst ecology clearly grew out of natural history and observational biology, quite independently of the experimental sciences, throughout the twentieth century it has been transformed, through its use of mathematics and systematisation, into the ultimate expression of the mechanical realist paradigm.[43] Henceforth, the totality of life on Earth is seen as an interconnected system of quantifiable states and motions – within which each and every living and non-living entity is a component described and identified in terms of its function within the totality. This finds its most grandiose expression in the holistic Gaia Hypothesis, as proposed by James Lovelock, in which ecology meets thermodynamic cybernetics, within which the mechanism of the thermodynamic chemical feedback loop becomes the dominant explanatory trope. The Earth becomes transformed into one gigantic machine of interconnected subsystems, which is to be understood as the totality of the technological framework of all the experimental sciences. The Earth becomes synonymous with the scientific world-picture. Within the technological framework, the value of any living body (whether an individual organism or a forest) can only be understood in terms of its instrumental value to the totality, in terms of its function and efficiency. The fact that it is far more complex than the clockwork universe imagined by Descartes and Newton does not change the representation of Earth as being fundamentally machine-like. Despite all its New Age pretensions, the Gaia Hypothesis is a refinement of the scientific world-picture, not a radical departure from it. It is as much a product of the technological framework of experimental science as was positivistic physics, chemistry, or genetics. Intrinsic value is a meaningless and antiquated concept within such a framework because something can only have value in terms of its function, as a mechanism or resource, within the total framework.

All measurement requires the selection, reduction, and specification of reference from within a technological framework that orders, defines, and relates the techniques and practices, as well as the set of possible outcomes, involved in setting up the experiment and making the measurements. The rise to dominance of biochemistry reveals the extent that techniques in the laboratory epistemologically eclipsed observation in the field. The behaviour on an animal in the wild is considered to be irrelevant for understanding the fundamental processes of life itself. Bringing together the techniques of physiology and chemistry though the innovation of a whole new generation of machines, premised upon the mechanisation of life itself, has reduced the study of how life begets life to the study of the chemistry of proteins and nucleic acids. The aim of this reduction is to explain the whole process of sexual reproduction and biological diversity in terms of the laws of chemistry and physics. It is premised on mechanical realism and technological innovation that emerged from the intersection of physics, chemistry, and physiology, yet is postulated as an explanation of evolution theory. Implicit in this transformation is the reduction of every living entity to those aspects that are responsive to technical interventions and that can be represented in terms of organic functions and natural mechanisms, reducing the complex and unpredictable organism to a simple set of machine performances. In short, biological science needs to technologically select, isolate, modify, and kill the organism *en masse*. In this way, for example, millions of rodents are transformed into a universal machine (termed as "the lab rat"), through selective breeding techniques, in order to produce a standard, reproducible, and transferable means of mapping out the limits and possibilities between the human interventions of the scientist and the machine performances of the animals' shared and repeatable responses to those interventions. The mechanisation of life has extended the jurisdiction of the scientific world-picture to include intelligence and consciousness. Henceforth, the study of the intelligence (in humans and other animals) has been reduced to the endocrinology of brains and nerves to such an extent that neuroscience emerged from the intersection between chemistry, computer science, and cybernetics rather than from behavioural biology or social psychology.[44] Mathematics is represented as a logical abstraction of our naturally evolved abilities to perceive and order information, and this representation situates science itself within the scientific world-picture as a consequence of natural selection.

Once the universality and inclusiveness of the scientific world-picture was accepted, then the development of scientific knowledge was reduced to unreflectively refining its content. Through the metaphysics of mechanical realism, modern science was founded upon the fundamental representation of mathematical rationalisations of machine performances as being the "neutral sign" for the existing natural order of things and how change occurs within that system. This metaphysical representation was able to represent itself as being a neutral representation of experience and, thus, was able to conceal itself as metaphysics. The positivistic interpretation of experimental science, as collecting the facts of experience in order to provide resources and tests for theory, became the naturalistic interpretation. The logic of discovery was one that necessarily ran apace with technological innovation, within which the properties of all matter was to be modelled by mechanistic functions that were tested in terms of their utility within the process of technological innovation, until, finally, even the understanding of the enquiring mind was to be mediated by technology and modelled as a sum of mechanisms. Nature is represented as quantifiable and objective in order to investigate it through the mathematical representation of the experiences of the machine performance of the technological apparatus. This paradigmatic representation of Nature has remained through all the paradigmatic shifts from Newtonian physics to relativistic and quantum physics, due to the enduring representation of the reality of that which is calculable and reproducible through the process of mathematically rationalising machine performances. Everything else is a matter of interpretation and all that changed during these so-called "scientific revolutions" was the set of interpretative representations to explain the performance of specific sets of machines. Thus, even though physics has utilised different interpretive metaphysics during its history, the operational metaphysics of mechanical realism has remained throughout. Scientific models are based upon a reified abstraction of their object in order to represent a set of machine performances as a natural process, within which the autonomy of the experimental apparatus in all its variations of performance can be represented as the consequence of natural mechanisms. Thus the machine can replace the natural phenomenon, preserving the empirical data, which can subsequently be interpreted using different theoretical representations. Bacon's new science was premised on the precepts of mechanical realism and represented true knowledge in terms of the alethic knowledge of the possibilities and limits of action and material practices. This knowledge was abstracted into "the laws of Nature". True knowledge does not consist in reasoned discourse (no matter how eloquent or revered the speaker may be), nor is it comprised of the conclusions of logical syllogisms, but it is based on the successful demonstration of the efficacy of an operation, procedure, or technique to effect some change in some particulars, discovering the workings of Nature or solving some practical problem. This reductionism has been metaphysically operational within the history of science since Locke proposed his notion of primary and secondary qualities as a clarification of the distinction between the subjective and objective that was proposed in Galileo's new science. This distinction is itself a product of the technological framework of the mathematical projection of the six simple machines over all phenomena. That which can be mechanised and mathematically abstracted is considered primary, objective, and part of non-human nature. That which cannot is considered secondary, subjective, and part of human nature. This duality resulted from a technological transformation of human thinking that transformed the limits of machines and mathematics into the boundaries of Nature and human psychology. The traditional Cartesian dualism between the objective and the subjective is a metaphorical de-centring of the relations between humans and machines in order to provide correspondence between labour and knowledge, permitting the subsequent removal of the machine from the account. This metaphysics leaves us with the writing on "the Book of Nature" while the scientific world-picture objectifies everything in such a way that it can be grasped as something available for use.

Metaphysics and modelling machine performances

As I have argued above, the scientific world-picture describes reality in terms of an integrated system of interconnected natural entities, mechanisms, and laws, disclosed through connecting componential machine performances and their associated theoretical representations. Mechanical realism permitted the form (logical and mathematical representations of mechanisms) and content (the facts of experience of machine performances) to be connected through the technological framework, while representing that connection in terms of "natural mechanisms" governed by "natural laws". This interpretation is not itself an empirical proposition. Hence, it is quite ironic that the positivistic turn, the rejection of metaphysics as being unscientific, was premised upon a conception of natural science that was only possible because of a metaphysical interpretation of the connection between the artificial and the natural. Positivism deals with the representations used in experiments to make, order, and interpret experience, as if they were logical abstractions of reality that exists independently of the human efforts to discover them. Consequently, positivism is based upon a metaphysical reification of a long history of technical struggles, judgements, and experimentation. This reification treats that history as if it were the immediate and neutral consequence of natural laws and the logical analysis of experience. Thus, once the development of the technological framework is completely removed from the account, then all experiments and theories could be represented as being separate from one another and unique confirmations of the same natural laws. It presents the mediated experiences – only possible due to a technical education on how to use and interpret scientific instruments and apparatus – as if they were unmediated first-person sensory experiences that either confirm or refute theoretical expectations. This is also the case with observational sciences (such as astronomy) within which the objectivity of observation depends on the use of learnt, technical interpretations and representations of both the objects and the instruments of observation. However, once we examine the role of technology within the experimental and observational sciences, we can see how technology objectively bridges this space between the observer and the observed, mediating those observations through representations and techniques. Thus all observations, in order to be objective observations, are based on skilled and theory-laden interpretations of how the technological apparatus and instruments work, and the empirical facts are determined by the historically developed technological means of experimentation and observation. The content of the facts of experience are determined from within the technological framework of the scientific research in question, and, in the case of novel research, making observations and taking measurements is only made possible by innovating refinements of the technological framework within which the experiments can be performed.

The neglect of the role of technology in mediating and constructing experience in the experimental sciences has led to a completely false consciousness about the meaning of experiment. The central characteristics of machine performances are uniformity, standardisation, and repeatability, and, therefore, the foremost scientific advantage is the amenability of machines to mathematical abstraction and rationalisation. Through technology, it is possible to transform any natural products into a set of constants and variables by first transforming them into machine performances. Hence, it should be of very little surprise that positivistic science provides increased predictive success and manipulative control. Once we examine how experimental science is done in practice, then we can see that the studies of the majority of the phenomena investigated by scientists involve the interpretation of the performances of machines and instruments. It is often the case that, when they are making observations, scientists only have direct experience of the numerical and analogue readings on calibrated meters, oscilloscopes, graph plotters, gauges, computer displays, and other instruments. The way that scientists describe and explain the performances of the experimental apparatus in terms of mechanisms and laws does not passively arise from their experience of the output of instrumentation. They need to interpret the performance of instruments in terms of models of how they work before they can make sense of these experiences and transform them into scientific observations, before they can be compared with theoretical expectations. Consequently, a close inspection of how scientists use models in experimentation shows that experimental science is not purely empirical, in the philosophical sense. Of course, it tests theories against experience, but it uses models in order to unify the particularities of the instrumentation into an observation of something, whether that something is a change of energy state, the expression of a gene, or whatever. The experimental scientist does not simply map out empirical regularities, such as when event A happens then event B follows, but uses causal accounts in order to provide definite measurements of the particularities of both event A and B, as well as techniques of intervention that are necessary for there to be an experiment at all. The scientist uses models, utilising representations of underlying mechanisms, to design, build, and operate instruments and apparatus, and, therefore, these models are implicit in subsequent uses and interpretations of these artificial devices to make further observations and theories. Once we recognise that every series of technical procedures and operations is represented as a sequence of procedural causes and effects, in order to make those operations meaningful, such as turning a dial to increase temperature or pressure, or increasing the intensity of an electromagnetic field, then we can see how observations and measurements of the responses to those interventions presuppose representations of the "causes" of experience in their models that are implicit to the design of the machines and instruments needed to make those observations and measurements in the first place. These models describe and explain the possibilities and actualities of the machine performances of the apparatus or instrument in terms of theoretical entities and mechanisms, and modern experimental science could not proceed without them.

Experimental scientists, whether or not they realise it, need to presuppose scientific realism at some level. The theoretical entities and mechanisms proposed as plausible explanations of observed phenomena are established as real through the modelling of instruments and apparatus as artificial devices capable of detecting the effects of human interventions in natural processes or states of affairs. The scientist assumes that accurate models in terms of functionality necessarily imply that they are accurate models of the phenomenon under investigation, and, consequently, these models disclose the underlying mechanisms of the phenomena under investigation.

Furthermore, a model is an essential link between mathematics and technology, providing the explanations needed to connect the quantified machine performances of calibrated apparatus and instrumentation with a "physical process". This is essential for quantified changes in the performance of apparatus to be related to variations in "physical processes" that are inferred from the response of the apparatus to human interventions. Models relate machine performances to "physical processes" in terms of visual, descriptive, explanatory, and intelligible representation of these changes. They are used to design and construct instruments, experiments, and computer simulations, as well as to calibrate measuring devices, and interpret experiences in novel contexts. Hence, in order to qualify as a legitimate model in experimental physics, any model has to facilitate the derivation of measurable quantities that can be connected to machine performances, but it also has to provide representations that make those machine performances intelligible as the consequences of "physical processes". Scientists will not consider any theory to be empirically adequate (or falsifiable) unless it has been (or can be) implemented within the technological context of ongoing experimental work. Examining how scientists use models to innovate experiments and observations reveals the complexity of the relationship between theory and practice in experimental science. Experimental observations are premised upon the legitimisation and justification of techniques; experiences that cannot be disclosed via communicable and publicly accepted techniques are excluded from being included as scientific. This constrains the scientific uses of any model to remain within the boundaries of manipulations and demonstrable context using publicly acceptable techniques, and changes in our ability to produce, represent, and explain experiences and practices involve the development of new models of what it is possible to cognate, represent, manipulate, and control. These models must be made public in the form of explanatory representations if they are to be successfully communicated and reproduced. These explanatory representations can take many forms, such as textual or verbal accounts, drawings or diagrams, apparatus or instruments, photographs, simulations, and models, and do not only take the form of mathematical theories. This means that the possibilities available for any particular model are limited by the scientists' accepted paradigm of what constitutes an intelligible explanation, as well as their expectations of what is measurable using the available technologies. This paradigmatic constraint situates the available choices of models to be consensually commensurable in relation to the already established models used within the scientific community. It is in relation to the common practices of the scientific community that the expectations of measurability and appropriateness are represented in terms of the functions of the available machines and expectations of what those machines can reasonably demonstrate. The technological context in which these representations are used acts as a constraint upon theory and practice, via a paradigmatic constraint upon the question of which models are acceptable and how they can be used, is central to the choice of which directions of experimentation are considered to be potentially fruitful.

Theoretical entities must be implemented in models of technological procedures and machine performances in order to be intelligible and available for experimentation, visualisation, and test.[45] Experimentation is an open-ended process of postulating models to link otherwise disparate theories and practices, and is not simply a process of comparing representations with observations. Representations are instrumentally required in order to make the interventions necessary for observations in the first instance.[46] Experimental science is not simply hypothesis testing or theory falsification, rather it is an ambiguous ongoing process of making representations of material practice in which observations, models, expectations, and techniques are developed simultaneously in the context of making the experiment work as an experiment within which different observers can agree that they are observing the same thing.[47] Theory and practice are linked by models that use explanatory representations of mechanisms that can be related to interventions through a series of implementations of techniques and the mediation of instruments and apparatus; therefore, theories are related to practices by using models to represent the relationship between human interventions and machine performances in terms of the realisation of natural mechanisms within the technological practices that build novel machines. These natural mechanisms are supposedly realised by technological operations that are interpreted using models that represent the interaction between the components of the machinery of the apparatus. By using models in this way, experimentation is taken to be the actual disclosure of theoretical entities that cause changes in the performance of the apparatus in response to human interventions. The ongoing work of experimentation is to provide models of the causes of experience that allow the objects and mechanisms of theories to become "observable" and "available" for manipulation by interventional and representational techniques. The only theoretical entities that are acceptable within experimental science (whether described as forces, fields, energy levels, or genes) are those that can be used to interpret machine performance, via a mechanistic model, and when these are instrumental in the innovation of novel machines they are taken to be real. The epistemological character of scientific knowledge is reduced to "know-how" and the study of natural phenomena is reduced to the search for fundamental mechanisms and their implementation in future scientific work. Experimental science is directed towards the achievement of a complete model of the experiment in terms of the mechanisms that supposedly govern the performances of apparatus. Such performances are taken to be the natural responses of the apparatus to human interventions and are taken by scientists to disclose natural mechanisms at work. Scientific knowledge has the characteristic form of a theoretical knowledge of how change happens due to the actualisation and exercise of natural mechanisms in accordance with natural law. Thus experimental science is premised upon the "how does it work?" question and it "tests" any answer to that question by attempting to produce and reproduce the proposed mechanisms in further technological innovation. Modelling, achieved through the explanatory connection of representations and techniques, allows a model to act as both *techne* and *episteme* by describing what something is in terms of how it works. Thus the thermodynamic properties of copper as a result of being subjected to an intense magnetic field, for example, are those, and only those, properties that can be measured and represented as a set of machine performances, in terms, such as magnetic resonance and damping, which are also represented as machine performances. Once these machine performances are taken to be the natural properties of copper, then the machine replaces the natural entity, which is henceforth understood in terms of a set of functions and explanatory representations, given in terms of mechanistic models. The natural entity has been replaced by a set of models juxtaposed with techniques and machine performances; the experimental apparatus is itself a metaphor for the natural phenomena under investigation. This show how there is an epistemological "gap" between the experiment and the natural world that can only be "bridged" using a metaphysical concept of "natural mechanism" to connect machine performances and the natural phenomena under investigation. By utilising such a concept, scientists cease to be working in the context of correspondence of theory and experience, but, instead, are working in the context of coherence between models of the functionality of the apparatus and models of the responses of the apparatus to human interventions.

Are the experimental sciences empirical?

Nancy Cartwright termed the very specific arrangements of objects and properties that have stable capacities or powers to generate empirical regularities as *nomological machines*.[48] Such a machine may be very simple, such as a rigid rod placed on a fulcrum that serves as a lever, or it may be very complicated, such as the Stanford Gravity Probe. The important feature of such machines, according to Cartwright, is that they possess capacities that generate regular behaviour when the machines are set running in the right conditions, shielded from unwanted outside causal influences that would interfere with its operation. Cartwright argued that capacities are more ontologically fundamental than empirical regularities because the existence of such regularities depends upon the existence of the capacities of a nomological machine. Laws only hold so long as such arrangements are in evidence and effectively shielded from interfering factors. Scientists construct models of the various capacities of the arrangement in specific circumstances, but in other circumstances different combinations of capacities may be evidence and a different law may be applicable. For example, in the absence of an empirically adequate model for the movements of a banknote in the wind, we have no empirically based reason to think that Newton's Second Law applies. If the situation is too complex to model then we cannot know whether this abstract law is relevant or there is a different law describing different capacities in that situation. The law and model are both empirically limited to the case of the nomological machine in question and, hence, we do not have any *a priori* reason to claim that the law applies beyond this machine (or set of machines). Cartwright demanded that, if we aim to be empirical, we should reject the validity of the fundamentalists' extension of the remit of the law to particular complicated situations for which we lack any empirically adequate model, and we should accept the patchwork and discontinuous nature of the Universe that is open to scientific investigation.

However, even though Cartwright raised some interesting critical points about the limits of the empirical claims about the applicability of natural laws that could be made on the basis of modelling machines, her argument neglected to address the metaphysical function of a concept of "natural mechanism" in extending and combining the models used in the laboratory to the wider and much messier natural world. Her interpretation of science is unable to explain the scientists' rationale for using nomological machines to model complex phenomena that they claim to pre-exist their efforts to model them. She did not explain how scientific fundamentalism was possible. However, once we take mechanical realism in account, then we can see how the function of a concept of "natural mechanism" is essential to connect the nomological machine with the complex phenomena of the natural world because, without it, we cannot explain the epistemological rationale of using machines to explore natural phenomena. As Roy Bhaskar argued, a concept of "mechanism" is essential to connect the closed system of the experiment with the open system of the natural world because, without such a concept, the classification of experimental physics as a natural science is completely unintelligible and arbitrary.[49] It is necessary to explain how experimenters connect their efforts in the artificial circumstances of the laboratory to the phenomena of the natural world, and, without it, scientists cannot explain how the empirical regularities were produced during an experiment, in terms of either tendencies or capacities, nor would they be able to extend those explanations to the natural world. Of course, models must be accurate in order to be widely acceptable, but if a theory does not explain natural phenomena, then it is an unsatisfactory theory because physicists seek to explain the existence of such phenomena in terms of the same causal accounts they develop and use to design, build, operate, interpret, and modify the machine performances that are intentionally produced to model those phenomena. The power of scientific models to provide intelligible explanations of phenomena in the complex real world is central to the whole process of establishing the intelligibility of experimental physics *qua* natural science, and it provides a conceptual continuity between the artificial world of the experiment and the natural world within which it is situated. Accordingly, the fact that working scientists are unable to predict the behaviour of natural phenomena, such as lightning, does not mean that they are unable to explain lightning in terms of the laws of electromagnetism, thermodynamics, mechanics, and optics. It is the explanatory power of models that leads physicists to be confident in the (approximate) correspondence of those models to underlying physical reality, while their predictive accuracy is limited to the much simpler apparatus used to model complex phenomena such as lightning. Physicists remain content to consider the phenomenon of a lightning storm to be explained even though they are unable to predict the behaviour of such a complex natural phenomenon. In fact, the "non-linear complexity" of such phenomena is itself postulated to explain their inability to accurately predict weather in general. Whereas Cartwright is quite correct to state that, if our understanding of the Universe is limited to the accurate models, the Universe is shown to be a discontinuous patchwork of such machines and that scientists should not treat natural laws as universal, but given that the experimental sciences transcend this limitation and explain phenomena that cannot be accurately modelled, all Cartwright has shown is that experimental sciences, such as physics, chemistry, or genetics, are not purely empirical sciences and that empiricism is an inadequate philosophy of science for these sciences.

Bhaskar implicitly accepted mechanical realism when he stated that things possess powers and liabilities in virtue of their internal structure of mechanisms.[50] His concepts of "power unrealised" and "power unexercised", to express the way that mechanisms can be suppressed by other mechanisms or when the conditions under which they are exercised are not right, in his claim that its microstructure or internal constitution determines the powers of a being, are founded on mechanical realism. Bhaskar identified power as belonging to the being in question and, thereby, asserted that this conception allowed him to identify the essence of a being in terms of its powers and limitations: "Dogs cannot fly or turn into stones, but they can move around the world and bark in all kinds of ways."[51] However, this conception of the essence of a dog is Aristotelian. Bhaskar attempted to simultaneously adopt an Aristotelian empirical analysis of the phenomenal characteristics of dogs in their natural environment *and* presuppose mechanical realism and define dogs in terms of their powers as defined by their internal structure. Aristotle defined a being's essence as being what something is in virtue of the matter of that being possessing particular characteristics. The identification of the essence of a being involved the empirical identification of which characteristics are essential in the sense of being causally fundamental to the identification of that thing as a member of a specific kind. This was not done in terms of a microstructure, or hidden reality, but was done according to the phenomenal appearance and behaviour of a being as distinct from other beings. The characteristics of barking, moving, four-leggedness, and so on were clustered as a multiplicity of characteristics for the purpose of classifying an individual as a dog. This involves an assertion of a set of normal canine characteristics within an environment that is taken to be natural for dogs. However, the dogs involved in the Soviet space program not only flew but also reached an orbit of the Earth. The dogs of the Palaeolithic on display in museums have petrified into stone. Dogs cannot bark in a vacuum or underwater. They cannot move around the world when they are caged or trapped in a box. Bhaskar, by associating barking and moving as the powers of dogs, was simultaneously asserting a normal environment for dogs to be in and a normal set of interactions within that environment. By doing this, he had not only presented an account of a normal existence for dogs, but by identifying the powers of a dog as being a consequence of its internal structure, he had neglected to attend to the way that power is emergent from within a complex of environmental relations and must be said to be a property of the relations. Changing its environment can change the powers of a dog; hence, the Aristotelian objection to experimentation is worthy of consideration, if we intend to base our scientific knowledge of the natural world upon experience. The behaviour of a dog in a laboratory does not reveal the nature of a dog in its natural environment. Consequently, my objection to Bhaskar is not that he had presumed that outer space is not a natural environment for a dog, but is the hidden appeal to mechanical realism that Bhaskar attempted to use to rhetorically move from dogs in their normal environment to a general account of power in terms of mechanisms. This arbitrarily asserts conventions and pre-empts the scientific inquiry he advocated.

If beings exercise and realise powers and liabilities in virtue of the contexts in which they are situated, then his attribution of powers to the beings themselves was arbitrary. If the attribution of powers and liabilities to beings is based upon their possible interaction, and possible outcomes, with other beings, then it was arbitrary to attribute powers to the beings themselves rather than the interaction between beings. If the powers are attributed to the interaction between beings, then neither "power unrealised" nor "power unexercised" say anything more than there is the absence of any empowering interaction. For example, the power of a hydrogen atom to combine with a chlorine atom, under suitable conditions, to produce a molecule of hydrochloric acid is not, as Bhaskar claimed, necessarily a property of the "internal electronic structure", but is, alternatively, a property of all the beings and relations involved in the interaction. It is an emergent property of the whole interaction. The identification of an "internal electronic structure" in fact depends on the total reality of the context within which the production and identification of a molecule of hydrochloric acid occurs. This will include the presence of hydrogen, chlorine, the suitable conditions for their combination, and interpretations of the whole technical process. Such an account will include all the conditions in virtue of which hydrogen and chlorine exist and are brought together. It is a power of the whole process of reproduction, from beginning to end, and the explanation of this process in terms of atomic structure is based upon an interpretation of this process. However, if we do not accept that interpretation, but locate power in the whole process instead, then hydrogen has no power to produce hydrochloric acid at all and is only an agent in this process. Thus we do not need a notion of "power unexercised" to describe the inability of hydrogen to produce hydrochloric acid when chlorine is absent. Nor do we need a notion of "power unrealised" to describe the inability of hydrogen and chlorine to produce hydrochloric acid if the conditions are unsuitable for the reaction. Hydrogen has no power, exercised or unexercised, realised or unrealised, in isolation, as an element. Given that hydrogen is only identifiable *qua* hydrogen in virtue of what it reproduces during certain kinds of interaction, it has no identity in isolation. By adopting a mechanical conception of essence, Bhaskar defined the is-ness of a thing in terms of what that thing does. Thus the being of any thing is de-limited by its productive agency in context, and the knowledge of a thing within a context of production is techneic rather than epistemic. Can we say that something without power, without identity, exists? Not on Bhaskar's mechanistic account.

In my view, this forced Bhaskar's hand. He rhetorically stepped outside his mechanistic account in two ways to preserve his mechanistic account. First, he defined an entity in terms of what it has the *potential* to do, and, secondly, he defined a set of normal conditions in which an entity would exercise its potential. However, by insisting that the potentials of entities were to be determined by experimental science, he has presented the closed system *as if it were the normal conditions* of an entity. Thus, for Bhaskar, how an entity is revealed by science is how the entity truly is, if it were not for the interference of other entities in the messy open system. However, this causes Bhaskar serious problems. If something, say hydrogen, is identified in terms of what it reproduces in certain kinds of interaction, and repeatability (as a deterministic regularity) only occurs within closed systems, as Bhaskar claimed, then without the notions of "power unexercised", or "power unrealised", theoretical objects, such as hydrogen atoms, cannot be said to exist in open systems at all. Thus, these notions depend upon mechanical realism in order to connect the closed and open systems. When we claim that we detect the presence of hydrogen in an open system, we have, in Bhaskar's terms, subjected the open system to closure by using the artificial closed system of the detection apparatus. The detection of hydrogen requires certain kinds of interaction to take the form of constant conjunctions. This may at first sight seem a strange interpretation, but the power of a hydrogen detection device to detect hydrogen in an open system is not simply a matter of hydrogen being present in the open system. It is characteristic of an interactive technological process of closure in terms of a deterministic confirmation or denial of the existence of hydrogen according to an anticipation of what the detectable properties of hydrogen are. This is itself a reproductive technological process. The interactive behaviour of hydrogen, its is-ness, is dependent upon what it interacts with. The identity of hydrogen is dependent upon a context in which hydrogen can only be said to be a participant and the term "hydrogen" is an index for a set of reproduced interactions, interventions, and performances within that context. It can only be said to have transfactuality to the extent that this context of reproduction has a shared set of agents. The contexts of using hydrogen in the production of hydrochloric acid and as the nuclear reactant in fusion bombs are taken to be independent because the processes of the industrial extraction and production of quantities of hydrogen, required by both chemical and nuclear utilisations of hydrogen, and the shared hydrogen identification instruments used to check purity, are not taken into account. The power to produce hydrochloric acid arises through the interaction of all these agents and cannot be isolated from the context of reproduction. It is a property of the technological labour processes that brings together and interacts these diverse agents. Once we take the shared production processes and identification instruments into account, then we can see how the so-called "independent contexts" are, in fact, historically and technologically related via techniques and instruments of reproduction. Thus the power to produce hydrochloric acid is a property of these technological processes, within which hydrogen and chlorine are agents, and, therefore, can only be understood by understanding the technological framework from which this power emerged as a power.

However, I am not saying that experimentation teaches us nothing about reality, but what I am saying is that how experience is constructed transforms what that experience is and what we understand about the object of experience. Experimental science transforms and mediates our understanding of Nature. Bhaskar's account of power, causes, and mechanisms in open systems may well be needed for a realist theory but it remains perpetually open to the criticism that it does not provide us with any infallible method by which we can identify which are necessary and which are apparent. There is no empirical basis for any such method. All such methods require metaphysical assumptions regarding the nature of the being and the world within which it exists. Thus, once we accept that experimental science is based upon mechanical realism, then we must acknowledge that it is not, in fact, an empirical and natural science. As argued above, experimental science is not limited to sensory experience but, rather, explores what is disclosed by means of sensory experience. However, Bhaskar was right to argue that, since its onset in sixteenth century, experimental science has had pretensions towards a "deeper" ontological relation than the mapping of temporal successions of empirical regularities. Sciences such as physics, chemistry, and genetics presuppose mechanical realism and utilise a concept of "natural mechanism" to connect the experiences produced in the laboratory to the events of the natural world outside. Even though the postulation of the existence of natural mechanisms are central to causal explanations of experience, when technologies are used to produce the context of experience, they cannot be brought into presence, as objects accessible to experience, but are heuristically essential to demonstrations, given in terms of the temporal sequences of human interventions and machine performances, in order to provide a causal account given in terms of the mechanisms that are proposed explain experience in terms of transcendent entities and structures. All the elements of the periodical table and all their compounds exhibit the character of artefacts because their representations cannot be separated from their properties, and therefore their nature, empirically determined through experimentation in order to explain and describe the functionality of such objects in terms of their underlying structure, is medicated by technical activity.[52] The properties of superfluids, the dynamics of phonons in crystals, the thermal capacities of metals, the properties of lasers, superconducting materials, solar neutrinos, the polarisation of light, and so on are all complex objects which are only disclosed through the mediation of machines, representations, and techniques. The establishment of scientific facts and theories about such objects requires putting techniques to work. The machines, theories, and techniques put to work to make investigation of these objects possible mediate experiences of these objects. The observational aspect of experimental work involves the active technical use (and modification) of theories, methods, and techniques. It is complex and there is not any possibility of being able to disentangle theories, techniques, and observations, except in hindsight through reconstruction. Without those techniques, representations, and machines, we would not experience these phenomena at all. The relationship between scientific experience and these otherwise invisible objects occurs by transforming machines and instruments into a means of disclosure of those phenomena by using theoretical representations and models. Such phenomena can only be understood in terms of the technological framework in which they are produced, but physicists are investigating these phenomena by using them as a means of disclosing the underlying causal mechanisms in operation in those machines in order to "test" their theoretical representations and models of those machine performances. All the fields of experimental physics, such as optics, electromagnetism, thermodynamics, quantum physics, solid-state physics, and even astrophysics, presuppose mechanical realism. The technological development of these fields of physics can be represented as a process of discovering the stratified ontology of Nature. Hence the mechanical properties can be explained by the properties of materials, which can be explained by the properties of chemicals, thermodynamic systems, and electromagnetism, which can be explained using quantum mechanical systems, and so on. The task of science is to explain this performativity in terms of "natural mechanisms". In order to understand how this task is successful we need to examine it as an ongoing historical process of trial-and-error efforts, within a society that values the practical uses of the prototypes produced during experimental research. The stable convergence and connection of different machines, via shared models and components, from within a historical process of technological innovation, allows the mechanisms postulated to explain the performance of these machines to be represent as a process of disclosing stratified ontological depth during the innovation of further machines within the same technological framework. In this sense, the ongoing innovation of the technological framework of experimental science can be represented as the ongoing refinement of the scientific world-picture. The technological framework achieves its autonomy because its history is suppressed.

The presupposition of mechanical realism allowed machines to become transparent means of disclosure at the service of "Man", used to disclose the causal principles of the Grand Machine, the Universe, and everything contained therein. This was taken to be a self-evidently rational exploration of the world, and, henceforth, notions of scientific progress have been implicitly premised upon appeals to technological innovation and power. This allowed human beings to justify the representation of the essence of reality in terms of an alien, mechanistic, material being that is only knowable through experimental science. It connected reason with objectivity through material and interventional practices, and our experiences of the alethic modalities of these experimental practices. Hence, the language of theoretical science and the empirical statement of the results of experiments both equate a being with its functions, and the definition of the being is determined by its material consequences in ongoing work. The facts of experience acquired through experimentation are based on a technical and often implicit understanding of function. In order to understand science, we need to avoid the tendency to abstract it into the discovery of facts about the natural world, but rather we need to examine it holistically within its history of production, its expected operation and function, its relations within its contents of application, and its actual relations within the wider world in which it is situated. Thus the walls of the scientific laboratory do not determine the definition of science. In other words, we need to examine science within its total reality, which is defined by the totality of uses and consequences. Accordingly, in order to provide a total definition of a science, we need to see both how it is transformed by society and, in turn, how it transforms society. Given that the total definition of science is understood in terms of these changing and specialised social relations, further changed and specialised as new results are brought into society, the total definition of science is perpetually incomplete and deferred until the future. Whether or not science can fulfil Bacon's dream has yet to be determined, and whether it is able to provide us with objective knowledge cannot be known until the whole societal project of modern science has explored every permutation of the innovation of the technological framework and the refinement of the scientific world-picture. It is in this sense that modern science is itself a grand societal experiment.

3
The Technological Society

The transformation from arts and crafts to modern science and technology arose primarily from a transformation in the conception of the origin of artificial and natural powers as having the same unitary origin and manifest according to the same principles or laws. The emergence of this metaphysical conceptual synthesis allowed modern experimental science to become possible as a means of using technology to discover natural mechanisms, while technology became represented as an artificial process of utilising the natural mechanisms discovered by experimental science. Nature could then be conceptualised in terms of universal mathematical laws, materials, mechanisms, functions, and efficient causes. This allowed the development of a representation of the human condition in terms of a scientific world-picture within which Nature provided the means for its own domination by Man, and technology was represented as a neutral process accessible to universal rationality, defined in terms of a technical rationality based on concepts of efficiency and productivity. Mechanical realism conceptually represents machines as the objective interface between Man and Nature, understanding Nature theoretically in terms of machine performances, situating Man within the scientific world-picture as one who grasps objective reality by using technology both to change his material conditions of existence and to refine the understanding of how Nature works through mapping out the contours between human intervention and machine performances. For the mechanical realist, human freedom does not consist in an independence from natural laws and mechanisms, but, rather, in the knowledge of natural laws and mechanisms and the possibility of making them dialectically work towards the definite end of increasing technological power. Ideally, this knowledge is to provide the Promethean promise of technologically increasing the possibility of liberating us from the organically evolved alethic modes of our "natural state of animality" and the arbitrary capriciousness of that state when it is one under constant threat of hunger, poverty, disease, squalor, fear, pain, ignorance, and subjugation to forces beyond human control. The experimental sciences have been organised within society according to their appropriateness and utility for the realisation of the values and ideals of society by providing novel machine prototypes and their associated powers.

Since the nineteenth-century Industrial Revolution, the experimental sciences became bound up with the development and application of technology to all spheres of life in order to solve the practical problems of wider society. It became a technoscience as it applied technology to the scientific systematisation, unification, and clarification of everything, in accordance with the practical demands of the civic, commercial, and military ambitions of the patrons of technoscientific researches. The technoscientific world-picture was exported and globalised through the educational and technical dissemination of European values, projects, technologies, scientists, and technicians. This Western colonial unification of the world in terms of a technoscientific mode of social organisation was totalitarian because the technological imperative towards efficiency absorbed, removed, and replaced any cultural plurality as it extended its grasp of the world in terms of a unification of all phenomena in order to maximise and rationalise power, co-ordination, and exploitation. This required the total organisation of humanity to maximise efficiency in every area of human endeavour. It was this totalitarianism that generated the monopoly of the technological imperative and made technological innovation an autonomous method of perpetually searching for the best technique to perform any given task. This created a universal mode of technical organisation that was internally directed towards its own destabilisation under the operation of the technological imperative to perpetually innovate more efficient means. Under the ideal of some perpetually deferred state of perfection, idealised in terms of absolute technological power, the whole of society was transformed into a process of realising this ideal, as a grand social experiment in the progressive realisation of manifest destiny: the construction of the technological society, free from the capriciousness of Nature, within which human freedom is manifested as unlimited power over our material conditions. In this sense, it is the whole technical construction of modern society as the technological society that is the experiment. This is why it is characteristic of a societal gamble. Thus, it must also be recognised that, according to the mechanical realist metaphysics within the societal gamble, the technologist and the physicist are engaged in the technological imperative to improve efficiency and productivity as a moral imperative.[1] The gathering and ordering of all plans and activity in a way that corresponds to technology, within a technological society, is seen to take us nearer to the realisation of the success of the societal gamble and the removal of the evils of the natural world. The drive to innovate, bringing novel inventions and new transformative powers into the world, is inherently a moral drive to construct the technological society. This imperative imposes a duty upon trained specialists to fulfil their social responsibility to discover and utilise the most efficient means at their disposal and to create further means to liberate their fellow human beings of the arbitrary capriciousness of the vulnerability, disease, ignorance, and premature death that Nature imposes upon us in our "natural state of animality". Within the mechanical realist conception of the technological society, human freedom is enhanced by the knowledge of natural law, and the technological imperative is inherently a moral imperative in which scientific truth is equated with the good. When freedom is associated with increased scientific knowledge and the productive powers, this entails the implicit goodness of the technological society and, consequently, the modern scientific society is premised upon a metaphysical conception of both its own possibility and the human good life.

Thus the productive possibilities of the experimental sciences are situated within the whole project of the construction of the technological society according to the estimation of the powers associated with each science. The further development of the powers, potentials, and possibilities of any science are, in their turn, shaped by the way that they are embedded and situated within the organisation of the work involved in the construction of the technological society. These potentials and possibilities should not be divorced from the organisation of the agencies and efforts of the labour processes that realise them. They are genuinely creative and transformative of the powers that they unleash upon the world. The organisation of the ongoing development and implementation of the experimental sciences has transformed the world in accordance with the posited anticipations of the form of truth and the human good life that are implicit to this project. The powers discovered by experimentation are the products of labour processes, situated within a historical trajectory, projected towards the discovery and liberation of power over our material conditions. Thus the experimental sciences innovate and produce their own creative transformations of reality as sets, clusters, and ensembles of machines, techniques, and powers, in accordance with an ontological moral equation that scientific truth is good. The technical causal accounts that are abstracted and communicated as understandings of technological powers in terms of natural mechanisms are further represented as being successfully tested during the successful reproduction of those labour processes and the cognition of their future possibilities in material practice. Scientific knowledge is abstracted in hindsight as a result of extending the closed system of the experiment and removing all hindrances to its reproduction, whilst transforming all contact with the natural world into series of causal explanations juxtaposed with clusters of machine performances and their associated powers. The truth of scientific knowledge is not only viewed in terms of its instrumental value for enhancing possibilities for producing an intelligible explanation of power, but is perpetually deferred to the future because its realisation is conceptually bound to its instrumental value in producing and explaining new powers, possibilities for labour, and machine prototypes. Thus the discoveries of the technosciences are inextricably bound to the efforts and decisions involved in the directions of the construction of the technological society, directed in accordance with civic, commercial, and military aspirations, while, as an ideal, the technological society remains an incomplete, imperfect, and ongoing experiment that is endlessly challenged to test and improve itself by innovating refinements of itself through the technosciences, embedded in the ongoing, experimental construction of an artificial world to replace the natural world, as a societal gamble on the superiority of this artificial world over the natural world.

The technological imperative

In *The Technological Society*, Ellul analysed technology in terms of its own reality, substance, and particular mode of being. He characterised the technological society as a perpetual state of social inequilibrium and defined "technique" as the totality of methods rationally arrived at and having absolute efficiency (for a given stage of development) in every field of human activity. "Technique" referred to any complex of standardised means for attaining any deliberate, stable, and rationalised productive behaviour or intentions to achieve predetermined results. It is the organised ensemble of practices that are used to achieve any goal, providing technical and quantitative solutions to technical problems, and it is objective in the sense that it is transmitted like a physical thing through the organisation of productive performances. Ellul argued that the situation could always differ from the contingent actuality of the present and he rejected any metaphysical notion of technological determinism. If technique is a "blind force" it is so because human beings have closed their eyes to the alternatives. It has the character of a technological imperative only because human beings respond to it as a demand upon us, bringing techniques to bear on all that is spontaneous, and systematically applying technical rationality in the division of labour, the creation of standards, the production of norms, and the reduction of method. Once we affirm that technical rationalisation of all aspects of life, then every intervention of technique demands the reduction of facts, forces, phenomena, means, and instruments to the logic of efficiency. The project of "the technical man" is a perpetual search for the "one best way" to achieve any designated goal and the perpetually expanding and irreversible role of technique is extended to all domains of life. Only that which is technical is considered to be part of civilisation and anything non-technical is excluded as inefficient, subjective, backward, irrational, or it is reduced to a technical form. The human subject becomes transformed into a rational agent, defined in terms of performance and function, as an integrated and articulated component, in an ensemble of functioning agents, within a totalitarian society obsessively driven towards the maximisation of efficiency, which, at any stage of achievement, objectively adopts the best available technique to achieve a given goal. Every best technique is made in reference to the satisfactory stabilisation of measurements, calculations, and productive practices, in relation to an intelligible causal account, emergent from the socio-technical consensus of expectations and estimations of alethic modality. It becomes another agent available for future work, ferociously tested by the technological imperative towards efficiency, as every agent is placed in competition with the others, to find the most efficient, socio-technical winner, as the one best way to achieve any given goal. Until it is replaced by another technique, the socio-technical winner, having achieved its autonomy in practice, due to the technically rational obligation to use it, remains indisputable as the efficient means and its use is no longer a matter for deliberation.

Ellul was concerned with the way that the dominance of the technological imperative towards efficiency has taken over all human activities and ordered them according to their utility, integrating and assimilating everything into the technological society, applying mechanisms, techniques, machine processes, and mechanistic logic to all areas of human activity. Technique has become autonomous as an integrated substance and medium of human agency within society.

The technological society is artificial, lacking spontaneity, opposed to Nature, and constructed through the societal gamble to create an artificial system that is supposedly more intelligible, controllable, and conducive to human well-being than the natural world. Just as hydroelectric installations take waterfalls and lead them into conduits, transforming them into electricity available for future work, so the technological society absorbs and appropriates the natural world. Ellul argued that the seventeenth and eighteenth centuries' scientific progress prepared the way for the nineteenth century's technical progress; scientific discoveries provided the necessary conditions but not the imperative.[2] The natural philosophies of the eighteenth century were concrete, bound up with material results, naturalistic, sought to know and exploit Nature, and constituted a preliminary phase of practical application. For Ellul, modern science became bound up with technique during the Industrial Revolution development and application of the machine and technique to all spheres of life. He accepted that the border between technological and scientific activities is not sharply defined, and that technique provides preparatory work for scientific synthesis and the creation of general explanatory theories. Science has become the instrument of technique because scientific discoveries are increasingly implemented in everyday life before the consequences of that implementation have been considered. Science requires the application of technique as a necessary condition of its existence; without technique science is merely hypothesis and theory. Using the example of Michael Faraday's early experimental work on the nature of matter, he argued that there is an increasing interaction between scientific work and technical preparation. He also used the example of the way that the steam engine was the product of technical trial-and-error sequences and scientific explanations came much later. Ellul observed that science has been becoming increasingly governed by technique since the nineteenth century to such an extent that the twentieth-century smashing of the atom and the smashing of Hiroshima (and Nagasaki) are manifestations of the same imperative.[3] He argued that the precision of any machine is only possible because of the elaboration of its design with mathematical rigour in accordance with its use to the extent that practical activity rejected gratuitous aesthetic preoccupations in favour of the idea that the most readily adaptable technique for use and mathematical calculation is taken to be the best. He accepted that this imperative somehow developed out of the science of mechanics but conceded that its origin was "mysterious and enigmatic" and possibly bound up with magical rituals.[4] For Ellul, the "modern worship of technique derives from man's ancestral worship of the mysterious and marvellous character of his own handiwork".[5]

According to Ellul, the dominance of technique upon capitalist economies imposes an impersonal centralism upon the economy through the mechanisms of the stock market, in which planning is reduced to "the order of the day" for the economy as a whole. This impersonalised centralism does not result "from the machinations of evil statesmen" and it cannot be controlled by public opinion. Ellul claimed that capitalism was only one aspect of the deep disorder of the nineteenth century and that capitalists were not the dominating motivating force behind the development of technique, but they were merely opportunistically aware of how to extract profit from technological developments.[6] Hence, the ills of technique – such as the flooding, pollution, and erosion caused by deforestation for industrial agriculture and the use of chemical fertilisers on the poor soil that remains – are products of the relentless development of technique to make agriculture more efficient, which brings obvious advantages as well as unforeseen consequences, and the capitalistic exploitation of cash crop markets is simply an opportunistic use of this development.[7] He argued that nineteenth-century liberal capitalism had become eclipsed by the efficiency of the technical method and the construction of a technological society.[8] Even the state, powerful that it is, has become dominated by the development of technique; the economy has become subordinate to this societal project.[9] Yet, Ellul recognised that capitalism opposes technique to such an extent that he interpreted the Marxist critique of capitalism as being essentially one that advocated the liberation of the development of the technological society (from the restraints of capitalism) because it is an essential good, which will free the proletariat and is the condition for the realisation of communism.[10] Ellul considered the Marxist vision of a communist system to be one orientated to technical progress, in all areas, giving free play to all technical automatism in every field of human activity, in order to achieve maximum efficiency, and, thereby, considered it to be the exemplar of the totalitarian technological society. He placed liberal capitalism in opposition to the construction of the technological society on the basis that the technological imperative is driven towards the achievement of maximum efficiency, whereas capitalism is driven to increasing profit.[11] Technical progress is antithetical to capitalistic individualism, which Ellul described as a form of anarchism, and he considered technique to be "the most important factor in the destruction of capitalism, much more than the revolt of the masses".[12] However, Ellul also recognised that

Capitalism checks technical progress that produces no profits; or that it promotes technical progress only in order to reserve for itself a monopoly... The pursuit of technical automatism would condemn capitalist enterprises to failure. The reaction of capitalism is well known: the patients of new machines are acquired and the machines are never put into operation.[13]

This view was also predominant in his later work. In *The Technological Bluff*, Ellul argued that the unthinking production and consumption of gadgets has nothing to do with efficiency.[14] Once we take all the resultants of a technique into account, including waste, pollution, external costs, excess production, and in-built obsolescence, then much of contemporary technological innovation has nothing to do with efficiency. Due to its reproductive power, modern technology has permitted the development of a universal economy of impersonal products, but the economic irony of the capitalist "market forces" model is that it is based upon an acceptance of the scarcity of resources, while it is profoundly wasteful and inefficient, exacerbating the problem of scarcity rather than alleviating it, and its foundation of consumerism is based on the creation of "need" rather than its satisfaction. This reveals a fundamental contradiction between capitalism's drive to profit and the technological society's drive to efficiency. What can we make of this? If capitalism is subordinate to the technological imperative, as Ellul claimed, and this imperative is antithetical to it, then how can capitalism resist and pervert it in this way? If the development of technique is the most destructive force over liberal capitalism, then why did the totalitarian, communist technological society, driven to maximise efficiency in all areas of human activity, fail to emerge as the dominant form of society?

Capitalism and the technological society

The collapse of feudalism and the emergence of capitalism was a consequence of the rising power of the Renaissance merchants.[15] The acquisitional ambition of mercantile commerce predates the development of industrial capitalism, but capitalism, in this form, was only able to emerge because it was able to acquire modern technology as its content. Industrial capitalism emerged from the intersection between feudalism, mercantile commerce, and modern technology. As Mumford argued, even though capitalism and modern technology conditioned and reacted to each other, they must be clearly distinguished from each other.[16] By the beginning of the nineteenth century, capitalism was the driving force in the dissemination of the practical new sciences and mechanisation. The incentive to mechanise all aspects of production lay in the greater profits that could be found through the enhanced productive power and efficiency of the machine. Mumford termed the technological complexes of the Industrial Revolution as the *paleotechnic phase*. He considered the railroad system to be the most efficient form of technics of the industrial complex produced during the paleotechnic phase. Inspired by the societal gamble, the scientists and inventors of the nineteenth century maintained the aim of achieving a more humane and rational society. The nineteenth century extended the Enlightenment effort to scientifically reorganise and rationalise every aspect of society, but due to the dominance of the paleotechnic phase, this extension was directed in accordance with the capitalist drive for profits rather than the technological imperatives. This effort was a continuation of the sixteenth-century humanistic preparatory stage of the scientific revolution, which Mumford termed as the *eotechnic phase*, but was empowered by the continuation of premodern ambitions for the increased wealth and power of a social elite (often over each other as well as the workers). At the onset of the modern era, the entrepreneurial spirit of capitalism empowered the technological society in order to develop the industrial capacities for precision engineering and mass production. The technological imperatives of maximising efficiency and productivity and the capitalist economic imperative of maximising profits were in phase, investors collaborated in supporting technological innovation and invention, which in turn fed their entrepreneurial zeal and provided them with increased wealth, but, when the capitalist economic imperative dominated technical rationality, while lagging behind the technological imperatives, investors withheld and suppressed technological innovation in order to maximise the profits on previous investments. This is apparent in the dominance of the coal, oil, and automobile industries and the suppression of alternative modes of electricity generation (such as wind, water, and solar) and public transportation (as well as electric cars). Such industries have sufficient resources and influence over governments to prevent the investment in and development of sustainable and alternative technologies, at least until their current investments have maximised their profitability and these industries can control and profit from those alternative technological innovations.[17] Hence, even though the nineteenth century was essentially an intensification of the unity between natural science and technology, efforts to invent and implement more efficient means of burning coal and transforming steam power into mechanical power were ignored by industrialists;

the Industrial Revolution was actually extremely inefficient because it squandered huge amounts of iron, coal, and human lives by continuing to implement and use inefficient technologies simply because they were more profitable in the short term.[18] The environmental pollution and a deterioration of human health during the nineteenth century were an inevitable (and, arguably, widely known) consequence of this inefficiency. However, iron, coal, and human beings were cheap and abundant. Thus, they were quite expendable. The dictates of the capitalist drive to maximise profits had sway over the technological imperative of maximising technical efficiency. The damaging effects that this policy had on human health and living conditions (including the illnesses and diseases caused by smog, pollution of air, soil, and rivers, overwork, low levels of sunlight, poor hygiene, overcrowding, the use of child labour, poor education, and the growth of slums) are historically documented as being commonplace during the nineteenth century. The destruction of varied human potential and social diversity – as well as forests and rivers – in order to maximise the profits and short-term gains of the industrial capitalists, offset the social value of the availability of cheap manufactured goods and the bulk transportation of commodities. Human beings were treated with the same disregard and exploitation as the natural world. They were simply a resource – a source of commodities – to be ordered, used, and discarded when exhausted, injured, or killed. The well-documented growth of child labour and the destruction of the craft base shows how skill levels and wages were reduced, while the length and intensity of working days increased, regardless of the social consequences, in order to reduce costs and maximise profits. The impoverishment and degradation of the factory workers was essential for the development of the whole basis of the Industrial Revolution as a capitalist revolution. Through the use of land monopoly laws and the destruction of traditional education for ordinary people, the industrial capitalists were able to nurture and propagate the foundations of industrial discipline. The factory system was the culmination of this industrial discipline and social engineering. Human beings were transformed into functions and components, and, after all alternatives had been destroyed (such as the crafts and rural agricultural base as modes of sustainable ways of life) and no other labour opportunities were present, these human beings were bound to machines because of their high levels of poverty, ignorance, and fear. The result of the Industrial Revolution was the creation and construction of an increasingly poor and restricted society for the majority of people, whilst a small minority grew extremely wealthy and powerful from the profits.

The paleotechnic phase was driven by an emergent capitalist economic imperative to maximise profits (by reducing costs), which transformed labour into a commodity, and utilised the machine as a mechanism through which the maximum surplus value could be extracted from human life. Out of sheer avarice and an overwhelming lust for power, combined with indifference (perhaps even contempt) for anything that could not be mechanised and utilised, the whole of social reality was transformed into a means of the production of wealth for a social elite, while representing the sacrifices of the workers as being inevitable consequences of social progress, even if such progress was illusionary for those that were making the sacrifices. As Mumford put it,

> The starvation of diminution of life was universal: a certain dullness and irresponsiveness, in short, a state of partial anaesthesia, became a condition of survival. At the very height of England's industrial squalor, when houses of the working classes were frequently built besides open sewers and when rows of them were being built back to back – at that very moment complacent scholars writing in the middle-class libraries could dwell upon the "filth" and "dirt" and "ignorance" of the Middle Ages, as compared with the enlightenment and cleanliness of their own.[19]

This inhumanity was considered to be morally justifiable because of the unquestioned social acceptance that technological innovation based on the experimental sciences was the route that humanity could overcome the savagery, brutality, squalor, and ignorance of his organic state and achieve the Enlightenment ideal of building a world within which science and technology would inevitably lead to a rational and humane society. Despite all the suffering that the Industrial Revolution was evidently causing, it was accepted as self-evidently true that progress was confirmed by each and every scientific discovery and technological innovation. This belief in progress was the unquestioningly acceptance of the societal gamble. It did not matter whether any new scientific discovery or technological innovation actually brought real benefits to society, or even whether it was evidently quite disastrous, because it was represented as a stepping stone towards the realisation of the ideal technological society within which all human needs would be satisfied. In short, each and every scientific discovery and technological innovation was a benefit to society simply because it had been achieved, regardless of any immediate benefits or harms it brought with it. On this basis, it simply did not matter that the industrial workers lived lives of drudgery as impoverished machine operators. It did not matter that the real quality of life of the workers of the Industrial Revolution was more impoverished than the medieval peasant. It did not matter that during the so-called Great War (1914–18) the new weapons of the industrial age (bombs, poison gas, artillery, flame throwers, chemical explosives, and the machine gun) killed more people and caused greater devastation than a century of warfare did in the medieval period. The facts of social reality simply did not matter. Progress was represented as the self-evident consequence of these terrible facts – which were represented as being the side effects of progress, as if they were simply an awful part of the process of human scientific and technological evolution. This was the historic mission of the construction of the technological society and the self-evident truth of that ideological abstract took positivistic precedence over the facts of concrete reality. As Mumford succinctly put it,

> Life was judged by the extent to which it ministered to progress, progress was not judged by the extent to which it ministered to life. The last possibility would have been fatal to admit: it would have transported the problem from the cosmic plane to a human one.[20]

The industrial and commercial ambitions of the capitalists required the representation of progress in terms of modern science, inventions, profits, power, machinery, luxuries, and comfort. These ideals were exported to other societies by means of trade, regardless of whether they actually benefited those societies, and were socially justified by allowing some of the benefits to the exploited, "underprivileged class" – as they were euphemistically called – provided that this was done prudently enough to keep the underprivileged class diligently at work in a state of passive and respectful submission.[21] The mechanisation of labour, developed by engineers and inventors in order to increase the profits of their paymasters, soon became beyond the control and comprehension of the factory worker. The operation upon the society of the emergent capitalist economic imperative reduced wages, lengthened working hours, deprived the workers with decent rest, recreation, and education, robbed children of a proper childhood and genuine opportunities for growth and self-determination, destroyed the centrality of family life, and left the worker ill, crippled, and impoverished in his old age (should he or she live that long).

The paleotechnic period, we have noted, was marked by the reckless waste of resources. Hot in the pursuit of immediate profits, the new exploiters gave no heed to the environment around them, nor to further consequences of their actions on the morrow. "What had posterity done for them?" In their haste, they over-reached themselves: they threw money into the rivers, let it escape in smoke in the air, handicapped themselves with their own litter and filth, prematurely exhausted the agricultural lands upon which they depended for food and fabrics.[22]

For Mumford, the paleotechnic phase was truly dreadful. It caused war, slavery, oppression, squalor, fear, disability, illness, and premature death on a massive scale. It was even more capricious than the Nature it supposedly countered, while it opposed, dominated, and perverted the technological imperative upon which its wealth was made possible. It was the product of the dominance of the capitalist economic imperative to increase efficiency (understood in terms of maximising profits by reducing costs) over the technological imperative to increase efficiency (understood in terms of optimising productivity by reducing effort required to achieve the work). Due to the high level of skill of premodern craftsmen, their exchange-value was based purely on the use-value, rarity, or beauty of their products, whereas labour was a commodity in the personhood of the slave. In modern industrial society, directed towards the mechanisation of skill and mass production, the craftsman continued to be a producer of luxury items and artworks for the wealthy, whereas the exchange-value of the worker became measured in terms of wages and time. The scientific world-picture, within which natural processes were represented in terms of mechanical efficiency, allowed the mechanical clock to represent the rational cosmos whilst ordering the working day based upon the regimentation of lived-world temporality (in terms of the standardisation, universalisation, unification, and quantification of time). Time became a commodity. It became equated with money either in terms of return on investment or wages for labour. The use of the clock to mechanise and regiment the working day reduced human existence to sequences and time-serving activities.[23] The mechanisation of labour involved a dehumanisation of the human being into sets and sequences of motions and the associated results, requiring a quantifiable form of social regimentation as its condition, appropriate to a technological framework within which that labour could be deskilled, controlled, and intensified in order to increase productivity while reducing wages. The capitalist economic imperative of maximising profits concentrated masses of workers in large factories, mechanising and standardising the processes of work, and reducing the standard of living for the workers to being that of bare sustenance. The individual worker was forced to accept the conditions of mechanised labour, starve, commit suicide, or become a criminal. Hence, even though Mumford was critical of Marx's assumption that machines have an autonomous essence, he agreed that capitalism had perverted technics, by placing it under the direction of the capitalist economic imperative that acts upon the machines at its disposal and transforms their operation in accordance with the lust for profit.[24] This lust opposes and perverts the emancipatory nature of technology, for the advantage of a few capitalists at the expense of many workers, and thus, if we agree with Marx that the purpose of any machine is to liberate human beings from labour, capitalism placed the operation of the machine in contradiction to its essence. However, according to Marx, it is because the capitalist considers the essence of the machine to be identical with its function within the capitalist production process that there cannot be any antagonism or contradiction, for the apologists for capitalism, between the nature of machinery and the capitalist employment of machinery. Thus, Marx's analysis shows the contradiction between capitalist economic imperative and the technological imperative, while, for the capitalist, the technological society is the means to achieve the capitalist economic imperative and therefore there cannot be any such contradiction. As Marx put it,

> The contradictions and antagonisms inseparable from the capitalist employment of machinery, do not exist, say [capitalist economists], since they do not arise out of the machinery, as such, but out of its capitalist employment! Since therefore machinery, considered alone shortens that hours of labour, but, when in the service of capital, lengthens them; since in itself it lightens labour, but when employed by capital, heightens the intensity of labour; since in itself it increases the wealth of the producers, but in the hands of capital, makes them paupers – for all these reasons and others besides, says the bourgeoisie economist without more ado, it is clear as noonday that all these contradictions are mere semblance of reality, and that, as a matter of fact, they neither have an actual nor a theoretical existence.[25]

Marx proposed a vision of a socialist society within which, through technology, everyone would be able to satisfy all material needs, while being sufficiently liberated from work in order to realise the creative and free essence of human labour, education, and arts. This vision was based upon a profound faith in the rationality of the societal gamble. Thus it was inherent to Marx's philosophy that once the proletariat controlled the means of production – the machines at their disposal – the emancipatory power of the machines would emancipate them. In other words, left to its own devices, developed under the technological imperative, the technological society would be a liberating and egalitarian society that would use the machines at its disposal to reduce the working day, lighten the intensity of labour, increase the wealth of the workers, and enhance the technical efficiency of society (resulting in less waste, pollution, overproduction, and unemployment). For Marx, emancipation is a historical event (not a mental act) that is brought about by changes in material conditions through the development of science and technology. It was a central feature of Marx's philosophy that labour is the mediator between society and Nature. He presupposed the validity of the precepts of mechanical realism, in which technology and labour are the material intermediaries between human beings and Nature, and social reality is transformed by labour – but it is the possibility and nature of that transformation that is taken to be objective, give human relations concrete meaning as productive relations and, hence, direct political theory towards the unification of the consciousness of the technological society. Thus Marx's theory assumed scientific realism in its formulation and characterisation of dialectics in order to inherently presuppose the rationality of the societal gamble, as it directs the form and content of the technological society. Dialectics cannot reveal anything final and absolute but, rather, reveals the transitory and contradictory character of everything. Marx proposed the dialectical method as being the only method by which it is possible to have knowledge of the totality, based on knowledge of the relations between all aspects of society and the society as a whole, as a historically evolving social reality. Its form and content are changed and developed during situated struggles to overcome social contradictions – not just in thought, but also in practice. While Marx did accept that social and economic relations of production were the products of history and not the laws of Nature, he did accept the objectivity of the laws of Nature and their effect on our organic and material conditions for existence. For Marx, human beings distinguish ourselves from other animals as soon as we begin to produce our means of subsistence, a step which is conditioned by our organic being, and by producing our means of subsistence we are indirectly producing our mode of existence. The possibility of dialectics depends on our ability to expound the real process of production, starting out from the material conditions of life itself, and to comprehend the forms of social inter-action connected with this and created by this mode of production. Thus, for Marx, the whole world is represented as a natural, historical, and intellectual process that is in constant motion, change, transform-ation, and development, and dialectics is the attempt to trace out the internal connection that makes a continuous whole of all this move-ment and development. Dialectical materialism was represented as the basis of all history and the true mode of analysis required to explain all the different theoretical and social products and modes of thought, and the whole of history – as a process – can be depicted in its totality, and therefore the interaction between various sides, aspects, and elements can be understood. Marxism is based on the conviction that the dialect-ical method is the scientific means to achieve progressive truth about the causes of social reality, only if it is developed and deepened by its advocates, in relation to the overarching project of the conscious construction of a better society, through scientifically guided productive activity understood within historical and political context. The form and content of dialectical materialism presuppose the societal gamble in the final goodness of the technological society. It is this conviction that locates Marx's theory – as the ideological expression of a method – within modernism, given that the objectivity of Nature is postulated as the ontological foundation for the historical and material development of society, providing humanity with the real possibility of change and mastery of the world. It is directed in accordance with the scientifically founded construction of the totality of social reality – the construc-tion of the universal socialist society emergent from industrialisation and modernisation of all aspects of human life – that makes dialect-ical materialism a revolutionary method, bound-up with the project of constructing and perfecting the technological society as a scientific society. History was represented as a unified process of the struggle for the development of the technological society; the completion and perfection of the technological society would be the end of history. This is evident from the implicit representation of human evolution within which human beings have the form of modern man, free creative labour is represented as the human essence, and the construction of the tech-nological society is represented as condition for human emancipation from our material conditions.

Marx considered natural science, as a theoretical relationship between human beings and Nature, to be internally bound together with prac-tical activity, industry, and with the development of labour, in order to disclose the real powers and potential of production. But, according to

Marx, due to the division of labour prevailing in class society, natural science is divorced from the material process of production and transformed into an abstraction of reality that was permeated with the ideology of capitalism. Whereas, for Marx, the human understanding of Nature was based upon practical activity, human beings were also considered to be a part of Nature for whom their identity emerges from the scientific understanding of Nature and the environment that human beings are able to construct, within the limits proscribed by Nature. The material practices of previous generations create and transform the aim and means of research, the instruments of observation, and the theoretical understanding of Nature, and the theories and practices of the natural sciences transformed by the productive and social conditions from which they emerge, due to its historically developing relations with commerce, industry, and human sensual activity. Accordingly, Marx considered Newtonian mechanics as one of the conditions for the development of industrial capitalism. But the relations between science and society are historical in character and dependent on development of production, which transforms the social conditions (such as class relations) that can assist or hinder the scientific development of production. Marx presupposed the objective material reality explored by natural sciences, but took great effort to examine the dialectical materialistic process through which an understanding of that objective material reality is produced. Thus, even though Marx accepted that the natural sciences discover natural mechanisms through a dialectical process of relating and developing theoretical and material practices, it is inherent to his theory that the conception of mechanism must change through this process. Thus, even though specific theories and practices are transitory parts of a dialectical process, it is through the engagement with reality through material practices that provides science with its objectivity and meaning. Even though Marx's conception of natural science was based upon an unreflective acceptance of mechanical realism, it would be a mistake to characterise his philosophical stance on the nature of the natural science as being deterministic or positivistic in any sense. While Marx considered natural science to be a transformative, revolutionary force, he was very much aware that the character and application of scientific methods and theories were historically shaped by their social and productive conditions. Thus the conception of natural science and the scientific world-picture were transformed with each and every major scientific discovery, and any real understanding of Nature is only possible on the basis of a developing understanding of the history of the development of human society, which forms a specific part of Nature. Conversely, the understanding of political economy also calls for the study of natural science as a condition of technical and economic development, as an essential condition for the development of productive power and efficiency. With all their qualitative differences, for Marx, natural and social sciences are unified in that they both study the relations between theory and practice within a material world from which they both historically emerge and change. Once society and production have evolved to a sufficiently scientific level of development, science will be transformed from a condition of the exploitation of the proletariat into a condition for the emancipation of humanity as a whole.

For Marx, the achievements of modern science have brought triumph to materialism, because they explain the objective laws, the material bases, the interconnection, the conditions, and causality of the material world, because these achievements enlarge the theoretical basis of the practical activity of Man directed towards the mastery of the forces of Nature. The ethical and aesthetic evolution of the consciousness of society is emergent as a unity between theory and practice, within the framework of the construction and perfection of the technological society developed in accordance with the technological imperatives to maximise efficiency and productivity. The conditions for the emergence and liberation of the consciousness of the proletariat – as being both a class and the basis of all productive power – are identical with the conditions for the unfettered construction of the technological society. The evolution of the class-consciousness of the proletariat is identical with the conscious development of the technological society, and the class struggle is the dialectical product of the contradiction between the capitalist economic imperatives and the technological imperatives within the societal construction of the technological society. However, capitalism places limits upon the development of science, once it has provided means to satisfy its commercial, military, and civic ambitions, because not only is further research expensive and requires long term and sustained investment, without any guarantee of profits, but increased productivity and efficiency can also place limits on the return on previous investments in the means of production. Hence, the crucial revolutionary liberation of the technological society was necessary to free the scientific development of the technological society from the perversions and limitations caused by capitalism. With the abolition of the feudal estates, capitalism could emerge as an economic imperative, imposing the form of society upon its technological content, whereas in the technological society, driven only by the technological imperative, both the form and content of society are technoscientific. Once the proletariat achieves class-consciousness, identifying itself as the productive power of society, then the demise of capitalism is inevitable once its basis for power turns against it. The workers' revolution is to be the spontaneous emergence of the awareness of the power of the proletariat – consciously identified with the technological society – and the obsolescence of the bourgeoisie and all class relations of ownership and control. The emerging class-consciousness of the proletariat is to be an aspiration towards the liberation of technological imperatives from the imposition of the structures of capitalist economics. It is the emergent consequence of the aspirations of the eotechnic phase – which have been perverted and jeopardised by the dominance of capitalism, which resists the completion of the technological society – which involves the transformation of all human beings into the proletariat in order for the form and content of society to be unified, and, therefore, dissolving the proletariat as a class. By dissolving all productive relations into the conscious totality of the technological society, the ideal of the socialist revolution is premised on the liberation of the technological society to become the totality that is conscious of itself as the totality. Thus the proletariat is the historical subject-object that circumscribes the form and content of the technological society, constituting the ontology upon which that society and its productivity are founded. It is implicit to Marxism (due to its uncritical acceptance of the rationality of the societal gamble) that the development of socialism is identical with the total integration of all relations into the teleology of the construction of the technological society, as a scientific society, free from the contradictions between the capitalist economic imperative to maximise profits and the technological imperatives to maximise efficiency and productivity. It is not only the case that, within Marxist theory, the essence of human being is free creative labour, but it is therefore absolutely imperative for the possibility of socialism that the teleology of human being and the technological society converge to the point of identity. In the technological society, intentionality and functionality are the same because human agency and labour are identical. Thus the abolishment of alienation is the conscious submission of the individual to the totality of the ontology of society, the form and content of which is the totality of the integrated productive and creative relations between human beings. Labour becomes a social category in which material dialectics becomes defined in terms of experimentation, innovation, and struggle for the emancipation of labour as a free, creative act that is performed for its utility within the social totality. The technological society totalises all being and becoming through the social category of labour to order everything within the teleological construction of itself. Thus, through material dialectics, labour becomes the subject-object of experience to such an extent that dialectics is manifest as the drive to the liberation, formulation, and perfection of itself as a universal method, as an expression of the liberation, formulation, and perfection of society. The Marxist definition of history as being the history of an unceasing overthrow of the objective realities that shape human life is a definition of modern history that is premised upon the societal gamble in the technological society, as being the means by which that overthrow could be possible. Socialism is the final culmination of the technological society in which both the revolution and "dictatorship of the proletariat" is the extension of the societal gamble into the totality of all human relations. The overcoming of alienation – the complete totality of the technological society – will dissolve individualism by evolving beyond it.

Marx's vision of a technoscientific socialist utopia was profoundly influential on Mumford characterisation of the *neotechnic phase* of the construction of the technological society in accordance with the scientific and technological imperative towards increased efficiency. He described this as being a definite counter-march against the paleotechnic methods because the role of modern science in the rational development of efficient methods, rather than trial and error, is the crucial aspect that distinguishes between the neotechnic from the paleotechnic phase. Thus he argued that the neotechnic phase began in the early nineteenth century with the scientific improvement of the efficiency of the water turbine, as well as the subsequent work on electromagnetism and thermodynamics. The paleotechnic phase of the construction and development of the technological society occurs when the capitalist economic imperative dominates the technological imperative and the construction of social reality. The neotechnic phase occurs when the technological imperatives freely operate upon the technological society, without hindrance from the capitalist economic imperatives (or any other influence). Both phases are equally totalitarian. Whilst the neotechnic phase is itself based on the scientific world-picture, bounded technical rationality, and the societal gamble, and it is directed, according to the technological imperative, to making the modern world universally better for all human beings, it does so only in so far as human beings are participants within the construction of the technological society. Thus the neotechnic phase extends itself to all aspects of life in order to improve them; it aims at producing better births and survival chances through technics (rather than simply increasing the number of births), providing better opportunities for health care and parental education, while removing illnesses, poverty, and the squalor of the paleotechnic phase.[26] Mumford argued that, in order to develop genuine technical rationality, transforming the productive and technological basis of society from the paleotechnic into the neotechnic phase, doing away with the wasteful and retarding effects of profiteering, we must devalue the capitalist economic imperative and integrate the scientific processes of technological innovation within a rational re-evaluation of human needs, to assimilate and co-ordinate technology in accordance with human needs and abilities. While the pecuniary interests of the paleotechnic phase dominate the mechanisation of labour, in order to rationalise the production process and reduce costs, then for the worker, labour is irrationalised because it is aesthetically reduced to a series of arbitrary procedures (that on an assembly line are dissociated from the final product) and arbitrarily quantified in terms of time spent for wages received. The productive process remains abstract and contrary to the interests and motivations of the workers, and, hence, they need to be coerced (through threats of unemployment) in order to participate. Given that coerced workers are either indifferent or antagonistic to the production process, this state of affairs inevitably leads to a reduction of worker efficiency, when measured in terms of productivity, even if it permits a reduction of labour costs. The reduction of the skill base of the general population, to justify the reduction of labour costs and increase the expendability of each worker, as part of the overall strategy of coercing labour, reduces the total social capacity for technological innovation and development. Whether at the level of the individual factory or the whole of society, there is a contradiction between the capitalist economic imperative and the technological imperatives, whereas an essential characteristic of the neotechnic phase is that it aims at creating a whole aesthetics of the technological society in order to improve the quality of life for all, by making the means of living life and overcoming problems more efficient in accordance with science. During the neotechnic phase, labour relations are organised in accordance with the technological imperatives to maximise efficiency – these relations are objectified and reified in terms of natural laws abstracted from the technological processes from which labour obtains its meaning as a rational activity.

However, as Mumford pointed out, the neotechnic phase had yet to develop its own form and organisation by the end of the 1920s, and, consequently, had not yet displaced the paleotechnic phase. The modern world is yet to embrace the neotechnic phase, and its social institutions lack the degree of adaptive and cooperative intelligence necessary for the fulfilment of the neotechnic phase of the technological society.

> Whereas the growth and multiplication of machines was a definite characteristic of the paleotechnic period, one may already say pretty confidently that the refinement, the diminution, and the partial elimination of the machine is a characteristic of the emerging neotechnic economy. The shrinkage of the machine to the provinces where its services are unique and indispensable is a necessary consequence of our better understanding of the machine itself and the world in which it functions.[27]

Due to the rise on consumerism, the middle class, and the small businesses in the industrialised world of the twentieth century, the class struggle of the so-called "developed world" has become transformed into a struggle against globalisation within the so-called "underdeveloped world". Through the operations of globalisation, the paleotechnic phase is as dominant in the modern world as it was when Mumford noted, over 75 years ago, that society was still undergoing a period of transition between the paleotechnic and neotechnic phase of modern civilisation. Again, this has led to the perversion of the technological imperatives into the creation of an increasingly technologically powerful war-machine directed in accordance with the dictates of the capitalist economic imperative, while the gap between the rich and the poor grows in developed countries, as increasing numbers of people in underdeveloped countries are compelled to work in industrialised factories and farms for their bare sustenance, in order to supply cheap manufactured goods to the markets of the developed countries, and the technical service needs of the developed countries are increasingly outsourced to developing countries with lower labour costs. Globalisation is a continuation of the paleotechnic phase. As Mumford put it,

> In the persistence of paleotechnic practices the original anti-vital bias of the machine is evident: Bellicose, money-centred, life-curbing, we continue to worship the twin deities, Mammon and Moloch, to say nothing of the more abysmally savage tribal gods... The neotechnic refinement of the machine, without a coordinate development of higher social purposes, has only magnified the possibilities of depravity and barbarism.[28]

I agree with Mumford that it is evident that paleotechnic phase dominates the politics and industry of the modern world, that there are still relentless nationalistic and class-based struggles, and that the natural and social environment is still being exploited to the point of total degradation without any regard for the long-term consequences. As Mumford put it, "the paleotechnic remains a barbarising influence. To deny this would be to cling to a fool's paradise".[29] The paleotechnic phase is still very much with us (perhaps, with increased globalisation, even more intensely today than in Mumford's time), and if the dominance of the methods and tactics that it has produced are not supplanted, then "the very basis of technics itself may be undermined, and our relapse into barbarism will go at a speed directly proportional to the complication and refinement of our present technological inheritance".[30] However, while I accept that the economic and technological imperatives must be analysed as distinct, often contradictory, imperatives that operate upon the construction of society, the history of the development of the modern world shows that economics cannot be divorced from technoscience, and that the scientific conception of the natural world, developed from the sixteenth century onwards, was a preparatory stage for the development of both the capitalist economic imperative and the technological imperative. Both the form and content of modern society are emergent from the contradictory interaction between capitalist economic imperative and the technological imperative, which becomes manifest as the dialectic between the paleotechnic and neotechnic phases of the construction of the technological society.

Mumford failed to recognise that the eotechnic phase, characterised by the Francis Bacon's dream of new discoveries and powers, to liberate human beings from the capriciousness of Nature and material constraint, had been directed and developed in accordance with civic, commercial, and military ambitions from the onset. Modern science and technology were never developed in isolation from the obsession with increased wealth and power, and, therefore, the scientific world-picture and the changing derivations of technical efficiency and aesthetics from its refinements are implicated in the paleotechnic phase from the onset. While the neotechnic phase does prioritise the technoscientific development of society, rather than using technology to increase the wealth of a social elite, it does not have the pure connection with an eotechnic phase that Mumford assumed the paleotechnic perverted. The paleotechnic was a development of the eotechnic, serving the extension of civic, mercantile, and military power, and, hence, the neotechnic is an idealisation that does not have any distinct historical connection with the eotechnic. Indeed, it remains an idealisation of the societal gamble – the construction of a perfect, technological society – but due to this lack of concrete independence from the content of the paleotechnic development of industry and technoscience, it remains an imaginary vision. Due to the dictates of the capitalist economic imperative upon the concrete and practical implementation of the technological imperative, capitalist economic aspirations have a transformative effect on both the form and content of bounded technical rationality, while being empowered by its application. All at stages of its construction, the technological society is ontologically based on the development of industrial capitalism, and it is for this reason that the neotechnic phase did not come to dominate over the paleotechnic, which continues today as relentless as during the nineteenth century, and the Marxist theoretical effort to synthesise a consciousness of the technological society, in order to dialectically and ontologically connect the rational form and technical content of society, ended up in the industrialisation of the Soviet Union as an intensification of the paleotechnic phase. Marxist theory is proposed as the ideological means to make the proletariat conscious of the extent that they are a unified class and that capitalism is by no means a natural and inevitable system for society – which is represented by capitalists as a necessary link in the chain of human development – but the truth is that it is nothing more than a contingent social power that has no inherent right to determine our actions. The empirical reality of the capitalist system is something that can be examined, challenged, and replaced with whatever form of society that human beings choose. The Marxist hope is that, through practical and critical activity based upon historical and scientific knowledge, human beings will consciously undertake this examination, challenge, and choose the form of society. However, even though Marx challenged the "natural" basis of society, he uncritically accepted the mechanical realist representation of the foundation of technology, and presupposed that technology has a neutral and autonomous essence. Thus, while Marx criticised the "naturalistic" interpretation of the social sciences, he accepted that the experimental sciences were, in fact, natural sciences. The scientific world-picture represents the operations of technology as being the process of exploiting the complex of interacting mechanisms at our disposal, treating Nature as something at our disposal, as a resource. Due to Marx's unquestioning acceptance of the emancipatory essence of science, the foundations of his socialist vision of the construction of the technological society entailed the very structures of domination and inequality that he tried to expose and overcome.[31] Marx presupposed that the ontological foundation of technical rationality is transformed into a social reality from a natural reality (as revealed by the natural and experimental sciences) and, thus, presupposed that the material foundation of the technological society is based upon the rational utilisation of natural forces – hence, Marxism fails to recognise and analyse the substantive capitalistic basis for the structures and trajectories of technical rationality, except in terms of its bearing on economic and productive reality, which is represented as a perversion. Despite the fact that Marxism is emergent as a political form of technological rationality, as an effort to unify the consciousness of that society in order to consciously and rationally construct the technological society, in accordance with the aspirations of the societal gamble, it fails to recognise that the ontological basis of the technological society is in fact artificial and contingent upon the same exploitative and totalitarian foundations as industrial capitalism. Even if the domination of the technological society by the capitalist economic imperative was removed, the proletariat would ontologically remain identical with the technological society operating under the same abstract and reified relations between human beings as circumscribed in accordance with efficiency defined in terms of the implementation of natural mechanisms as discovered by the experimental sciences. The totalitarian ideal of the technological society is to order all of its mechanisms, components, and resources (including human beings and the natural world) and coherently integrate them into a complete and unified automaton. Thus the dialectical relationship between necessity and freedom, in opposition to and defined by an objective natural state of being, would remain represented as quantifiable and mechanised in terms of increased productivity, efficiency, and power; thus, the project of constructing socialism through industrialised communism was doomed to repeat the same inequalities and dominating structures as the capitalist societies that it was supposed to replace. Thus the Soviet Union was doomed to repeat and intensify the oppressiveness of the paleotechnic aspect of the technological society, once, after the Civil War, under the dictates of Stalin, it moved beyond the intensely complex, democratic "heroic period" into the centralised program of industrialisation. The "dictatorship of the proletariat" became a repetition and intensification of the capitalist industrial–military complex, which repeated and intensified all the injustice, horrors, brutality, and cruelties of nineteenth-century industrialisation, alongside Stalin's betrayal of the revolution and his construction of a corrupt, bureaucratic, centralised dictatorship that silenced or murdered all dissenting voices.

In many respects, with the hindsight of history, due to the mechanical realist precepts that underwrote both capitalist and Marxist economic theories, some degree of fatalism, equivocation, and recapitulation regarding the power of capitalism should be expected since the collapse of the Soviet Union. However, the view that capitalism is the unlimited and invincible victor of the twentieth century is one that blindly neglects to recognise and attend to the environmental unsustainability and social injustice of unchecked capitalism. The failure of socialism to establish itself as a genuine alternative to capitalism does not undermine Marx's (and Mumford's) criticisms of the inefficiencies and injustices of the capitalist society that the socialist vision was supposed to replace.[32] Even though the Industrial Revolution was premised upon the promise to produce goods for masses of people, which previously were available only for monarchs, the feudal nobility, and wealthy merchants, while improvements in technology promised to eliminate the barriers to universally shared wealth, it is evident from the position of historical hindsight that industrialisation produced greater inequality and concentration of wealth and power in the hands of a social elite. As Marx argued, the activities of capitalists are not just simply parts of an economic process of maximising surplus value, but are perpetually driven to reproduce the class structure.[33] The capitalist economy is a consciously driven social process that is directed to maintaining the autonomy of the capitalists as a social elite, directed to perpetuate the proletariat as the instrument of capital, as well as reproduce the proletariat acceptance of capitalist propaganda regarding the intellectual, moral, and natural superiority of the capitalists. An analysis of the class relations between the owners of production and the exploited workers, who actually produce the wealth and productive power, is central to any rational and moral development of technoscience to the extent that the resolution of this inequality is a precondition of rationally and morally using technoscience to solve human problems in a way that equally benefits all humanity. Technological power and labour converge as they dialectically transform one into a manifestation of the other, and this convergence establishes the ontological basis of the technological society. However, the operation of the capitalist economic imperative disrupts that convergence in order to represent the dialectic between technological power and labour, as being governed by a "natural" master–slave dialectic within which the efficient is defined in terms of reduced costs rather than the scientific application of natural mechanisms. Thus efficiency is defined by the devaluation of labour rather than the empowerment of productivity. Herein lies the source of contradiction between capitalism and the technological society. Capitalism is ontologically founded upon the technological society, but the capitalist economic imperative to maximise profits and the technological imperatives to maximise efficiency and productivity only ran in phase during the eighteenth and early nineteenth century. During this period, the economic imperative empowered the technological imperatives, whilst they in turn provided the means to reduce costs and increase profits. However, from the mid-nineteenth century onwards, due to the scientific development of the technological society, the technological imperatives began to outpace the capitalist economic imperatives in terms of efficiency and productivity. It became possible to provide longer lasting and precise manufactured goods at a lower cost; hence capitalist economic imperatives required artificial reductions of quality, in-built obsolescence, inflations of price, and the use of antiquated industrial processes in order to maximise profits. The divergence between these imperatives became the contradiction at the heart of the modern world which traded-off universal technological progress for increased profits for a few. It is the failure of the capitalist economic system to integrate and equate visions of society with the technological imperatives, in order to affirm the effort to realise that vision as a moral imperative, which has been its ideological failure, and the resultant failure of the modern world to satisfy the societal gamble by constructing a world that is the best of all possible worlds. It is due to this failure that the possibility of a descent into a new and terrible form of barbarism looms over the world.

One-dimensional man and the societal gamble

In *One-Dimensional Man*, Herbert Marcuse presented a social critique of the structures of contemporary industrial society as a dominating society that represses human individuality, aspirations, ideas, and values that cannot be defined in terms of the narrow bounds of technical rationality.[34] Marcuse described this society as a technological society within which production, labour, consumption, leisure, culture, and thought are integrated into the dominating structures of that society in order to produce an advanced state of conformity. Like Ellul, he considered this technological society to be a profound threat to human individuality and liberation. Also, like Ellul, he considered technology in terms of a broad, social definition that did not define it merely as the sum-total of tools, instruments, and material practices, which can be isolated from its social and political contexts of emergence, but, instead, defined it in terms of a social system that, as a totality, determines the productive activity, needs, and aspirations of the individual, as well as the socially needed occupations, skills, attitudes, and the operations of servicing and extending machines. Technology cannot be isolated from the uses that it is put to. It mediates the thought and action of the individual in the technological society, negating the distinction between public and private life, between individual and social needs, and institutes an effective form of social control and cohesion through conformity. Thoughts and actions that conform to the operations of the technological society are brought together and empowered, whereas oppositional actions are prevented and corresponding thoughts are left speculative and impotent. The technological society is a totalitarian system of domination – in both industrial capitalist and communist countries – within which the operation of the system pervades each and every thought and action. Thus political rationality becomes reduced to technical rationality in industrial societies. As Marcuse argued, the totalitarianism of the technological society is not enforced by any tyrant or police state, but by the overwhelming and anonymous technological power and efficiency of that society.[35] Political transformations are themselves reduced to technological refinements and innovations, and, thus, each and every qualitative social change and political act is directed into the technological development of the technological society.

Marcuse described how technical rationality was implicated in the construction of a system of totalitarian social control and domination. Within the capitalist-driven construction of the technological society, individual human beings are increasingly integrated into an economic system that demands the total accommodation and submission of all human beings into that system. Increasingly, individual human beings are compelled to conform to the norms and practices of society. The structures of bureaucracy, planning, and management within industrial capitalist society have created a totally administered society without opposition that threatens individuality because it destroys all possibilities of radical social change. The technological society orders and controls human beings by integrating them into a system of organising all thought and action in such a way as to only allow change and reason to operate within its sanctioned institutions and frameworks. Once alternatives become impossible then society becomes "one-dimensional". Marcuse recognised that technical progress is defined by the movement towards ameliorating the human condition as its goal, and this goal was rooted in using "the scientific conquest of Nature for the scientific conquest of man" and was able to defeat all protest in the name of the historical prospects of "freedom from toil and domination" in order to abolish labour and struggle (between human beings, as well as against Nature) to achieve "the pacification of existence". He argued that once the technological society became capable of realising the liberation of human beings from toil and struggle, it closed itself off from this possibility, and became based on a principle of containment of technical progress in order to preserve the capitalist vested interests in maintaining control through owning the means to alleviate immediate scarcity and need. Hence, Marcuse argued that industrial capitalist society is driven by

> a trend towards the consummation of technological rationality, and intense efforts to contain this trend within the established institutions. Here is the internal contradiction of this civilization: the irrational element in its rationality.[36]

Hence, he argued that the claim that the capitalist economic system is rational is based on a conception of the "economic prosperity" of industrial capitalism, which ignores the waste, violence, destruction, exploitation, and repression upon which that system is sustained. The self-proclaimed rationality of the advocates and apologists for the capitalist economic system is itself irrational. He was highly critical of the way that "democracy" in capitalist society is based on media manipulation and conformity; how its wealth is generated through a dehumanising, alienating, slave labour system; the ideological fetishism involved in consumerism; and the insanity of the highly dangerous military–industrial complex upon which advanced capitalism depends. Without putting it in these terms, Marcuse argued that there is an internal contradiction between the capitalist economic imperatives and the technological imperatives. Like Mumford, Marcuse seems to equate the technological imperatives (when unfettered or uncontained by the capitalist economic imperative) with the Marxist vision of an industrial socialist society. Through political revolution, the political apparatus of capitalism and the contradiction of private ownership of the means of production and social needs are to be destroyed, technical rationality is to be freed from the irrational constraints and contradictions of capitalism, and will thus be able to sustain and consummate itself in the new socialist society. Once the control of the technological society will be placed in the hands of the workers, engineers, technicians, and scientists, who will take control as a matter of survival, then there would be a qualitative change in the nature of the technological society from satisfying the needs of a small social elite to satisfying the needs of the technological society itself.[37]

Marcuse presupposed that human beings possess the ability to make a true distinction between real and illusionary needs. Once this distinction is made, then it becomes possible to show that the contemporary media representations of "individualism" and "freedom", as well as the postulated means to achieve them (i.e. consumerism and the acquisition of capital), are actually dominating forms of propaganda, from which the individual needs to be liberated in order to preserve his or her individuality. Even though the economic and political structures of contemporary "democracies" are represented as being the condition of progress and liberty, they are in reality systematised mechanisms of domination that coerce individuals to conform to societal norms and practices, strengthening them, and perpetuating the very system that enslaves us. Once the individual's economic activity is limited to sell his or her labour as a commodity in accordance with the dictates of a system over which he or she has no influence or control, the individual becomes a functionary of that system. Once the individual's political activity is limited to vote for the choice of representative from between members of the same social elite, the "democratic" participation of the individual is reduced to being a functionary for the ratification of the illusion that the system is democratic. Once the media is under the control of a social elite, which can control and suppress information and the voices of genuine opposition, while being an instrument for "shaping public opinion", the media becomes nothing more than a relentless propaganda machine to establish conformity to the norms and practices of society. Once all the individual members of society are completely integrated into the system, as a system of conforming functionaries, then society is able to absorb and dissolve all opposition. It is thus able to direct all thought and action indefinitely, perpetually stabilising capitalism, on the basis of being the champion of human freedom, happiness, and individuality, regardless of its destruction of the environment, dependence on a globalised war-machine dedicated to perpetually create conflict and fear, while eroding all individuality and resistance, in order to sustain an increasingly wealthy social elite at the expense of the majority of the world's population. Hence, the "stability" of the capitalist system is only possible because it is able to conceal its contradictions and reality by presenting and disseminating the illusion of its own rationality, necessity, and beneficence, by creating the problems that it postulates itself as the solution for. However, it is the reality of these contradictions that undermine the system's stability and expose the extent that the stability of capitalism is an illusion propagated through mass deception and domination.

Marcuse considered technical rationality to be enframing all aspects of human life, imposing technological imperatives upon all forms of though and action, in such a dominating mode of instrumentalism that it ultimately erodes human freedom and genuine individuality. Technical rationality uncritically conforms to existing norms and practices, and, therefore, cannot provide the basis for critical and rational evaluation of societal norms and practices, except in the narrow sense of examining the instrumental efficiency of practices in relation to norms. As an alternative to this narrow mode of rationality, Marcuse attempted to develop a critical and dialectical mode of thinking, which was to negate existing forms and categories of thought in favour of realising higher potentialities, norms, and categories of reasoning. This involved the development of the human ability to abstract universal concepts and categories, instead of merely conforming to the concepts and categories given by society, in order to create a critical standpoint from which the conditions that suppress the potential for self-determination could be negated. Marcuse advocated the critical development of historical analysis of these universal concepts as the means to critique and transcend the empirical and its validation of ordinary discourse and behaviour. Critical and dialectical social philosophy was to theoretically facilitate the processes of genuine social resistance and change by critically establishing the basis for individual transformation in relation to the ideals and vision of liberation and happiness, against which current tendencies and conditions could be critically evaluated and rejected in relation to the potential implied by the application of our universal concepts, such as "freedom" and "happiness", and the way that empirical actuality does not achieve these potentials. For Marcuse, scientific thought is unable to place concepts in opposition to the facts of immediate experience, and, hence, it is unable to sustain the dialectical tension between "is" and "ought", despite the fact that science establishes its truth against experience. Science must conform to experience and therefore cannot make a judgement that condemns the established reality. A new scientific concept can only be opposed to the previous scientific concept to the extent that it conforms to experience better, but any conceptual opposition to experience and the current state of affairs is taken by science to be a sign of falsehood.

According to Marcuse, the aim of critical social theory is one of providing the theoretical basis for radical social change. The theoretical task is to show the historical contingency of contemporary norms and practices, while also showing that there are real possibilities for alternative norms and practices, based on a rigorously historical analysis of social tendencies in theory and practice. However, Marcuse presupposed that it is possible to correctly make distinctions between existence and essence, actuality and potentiality, and appearance and reality, by appealing to universal concepts. Marcuse was inspired by the Renaissance struggle against superstition, irrationality, and dogma, as expressed in both art and philosophy.[38] He considered it to be the creative source for individual rationality and liberation, upon which the real creation of an advanced and rational civilisation is to be founded, and technical rationality undermines and suppresses this creative source. By positing an idealisation of human essence and rationality, norms and potentials for human happiness and freedom could be identified and articulated, which could then be opposed to and negate the existing state of affairs and all its illusionary and irrational values and norms. Marcuse presupposed that the essence of the human subject is that of a free, creative, and self-determining being, which contains possibilities to be realised and qualities, such as values, aesthetic characteristics, and aspirations, standing in metaphysical opposition to an object-world, available for the cultivation and enhancement of human life. Reason links the subject and the object-world, placing the two in dialectical opposition, as a metaphysical duality, which allows reason to intuit the essence of reality to the subject in mediated contradiction to appearances. This essence is increasingly suppressed by the technological society, within which aspirations, values, and needs are represented as being objective norms for which there are objective means to satisfy them. He argued that Renaissance metaphysics has been superseded by technology that replaces the metaphysical and sublimating concepts of subjectivity and objectivity with practical and desublimating concepts of instrumentality and efficacy, which are simply understood through a common principle of ordering all thought and action in accordance with the arrangement and sequencing of means and ends. He argued that the dialectical relation between subject and object, itself a precondition for the subject's ability to negate the current state of existence and realise potentials and possibilities that do not exist yet, is eroded within the technological society because it assimilates the subject within the object-world. All relations between the subject and the object-world have become mediated and defined by the technological society, thus the human ability to critically identify real needs and engage in authentic, original activities to satisfy them has been eroded. Thus, "one-dimensional man" has his or her intentions and means to achieve them presented to him or her, his or her capacity for authentic thought and action is eroded and forgotten, and, finally, he or she has no alternative but to submit to the imperatives of the technological society and conform to its norms and practices. Once human beings have conformed to the technological imperative, then we become objects for manipulation, administration, management, and control. Hence, we become alienated from our essence, our free and creative subjectivity, and are no longer capable of realising our potential to live authentic, self-determined, social and economic lives. We lose our individuality as soon as we submit to becoming another object within the technological framework of the administration of society. We not only lose our ability to transform our existence, but we lose our ability to even comprehend it, and, therefore, the technological society atrophies the possibilities of genuine social change and human emancipation.

Once the subject has been assimilated, it becomes subjected to the same technical norms and practices as any other object, and, hence, loses the recognition of the distinctly rational ability to engage in transforming norms and practices in order to discover and realise more liberating possibilities. However, he failed to realise the extent that technical progress is rooted in a moral and metaphysical project that is profoundly embedded in the psychology and ideology of the Enlightenment ideals. However, once we recognise the foundation of the technological society upon the metaphysics of mechanical realism, the scientific world-picture, and the societal gamble, we can recognise that the technological society is not in opposition to Renaissance metaphysics, but, in fact, is a consequence of it. Marcuse was aware that modern natural science "develops under the *technological a priori* which projects nature as potential instrumentality, stuff of control and organization".[39] He recognised that this "technological *a priori*" involved a transformation of both society and Nature, which is understood as the *a priori* intuition of the Universe as a technological reality within which science projects and responds. Matter is understood, in terms of pure science, not simply in terms of its practical use for a specific task, but as instrumentality in itself. Hence, even though the rationality of pure science is not immediately directed to specific practical ends, it is itself a form of technical operationalism that is developed under an instrumentalist horizon.[40]

The principles of modern science were *a priori* structured in such a way that they could serve as conceptual instruments for a universe of self-propelling, productive control; theoretical operationalism came to correspond to practical operationalism. The scientific method which led to the ever-more-effective domination of nature thus came to provide the pure concepts as well as the instrumentalities for the ever-more-effective domination of man by man *through* the domination of nature. Theoretical reason, remaining pure and neutral, entered into the service of practical reason. The merger proved beneficial to both. Today, domination perpetuates and extends itself not only through technology but *as* technology, and the latter provides the great legitimation of the expanding political power that absorbs all spheres of culture.[41]

It was for this reason that Marcuse considered scientific rationality to be a mode of political rationality that was bound up with technology, as a mode of social control and domination, due to its function as a

rationalization of the unfreedom of man and demonstrates the 'technical' impossibility of being autonomous, of determining one's own life. For this unfreedom appears neither as irrational nor as political, but rather as submission to the technical apparatus which enlarges the comforts of life and increases the productivity of labour.[42]

However, due to his neglect to attend to the question of how it was possible for Nature to be ontologically represented as the technological *a priori* and for a technological science to be epistemologically represented as natural philosophy, since the Renaissance, he failed to recognise the shared metaphysical precepts that made modern science and technology both possible as aspects of each other.[43] No doubt due to the influence of Husserl and Heidegger, Marcuse treated scientific thought as if it were positivistic.[44] However, as I have already argued, the characterisation of scientific thought as positivistic has neglected to attend to the metaphysical basis of experimentation that permitted the epistemological and ontological aspects of science and technology to be represented as being consequences of the same causality.

The technological society is founded upon the metaphysical postulation of the emerging ontological goodness of the technological society because it is the scientific means by which human liberation and progress become possible. It is itself a negation of the human organic and natural state of animality, with all the ignobility, suffering, and limitation that state entails. It is a dialectical confrontation with Nature, postulated as movement to construct a replacement and improvement upon Nature, driven towards its own conception of perfection as absolute human power and freedom from all limitation and constraint. Due to his neglect of this metaphysical foundation, Marcuse did not recognise that the technological society is also based on an equation between truth and value. In fact, for the technosciences, a valueless truth is an untestable contradiction in terms, but the value of truth is understood in terms of its value for the discovery of more truth and power, rather than simply having immediate practical use. Marcuse was quite correct about that, but he failed to reveal the cultural meanings that permit technoscience to be represented within the modern world as the source of human liberation and enlightenment. Without addressing how this representation was possible, his critique of the technological society, although quite correct about its outward appearance, failed to expose and critically analyse the metaphysical foundations of that society upon the societal gamble. Marcuse was only able to address the social consequences of the societal gamble, rather than examine the historical conditions that made its representation as a rational gamble possible. However, once we recognise this metaphysical foundation, we can recognise that the technological society is itself incomplete and is an idealisation of the final goal of all struggle, labour, science, and technological innovation. Hence, it is not a rejection of the classical dialectic between Being and Non-Being at all, as Marcuse claimed,[45] and, hence, the technological society has not answered "the question of Being", but is actually a process of answering this question by creating itself as the answer to this question. The technological society is an idealisation of the future of the whole process of innovating itself from the past, concrete, technological ontology created in confrontation with Nature, and thus is bound up with its own Becoming. It is at this level that the reified, abstract potential of the creation of the technological society is idealised as the Being of Becoming, to be realised by acting in accordance with natural law and implementing natural mechanisms in the ongoing construction and perfection of society. The technological society is postulated upon its conception as the completion of history, the destiny of the modern world, and a negation of the past and the natural world. The process of creating the technological society as a replacement for the natural world is itself the modern dialectic between Being and Non-Being that is made immediate and concrete in its perpetual Becoming. The technological society, driven by the technological and capitalist economic imperatives, is inherently premised upon a dialectical movement from the past into the future (transforming the past into the increased potentiality of wealth and power to be realised in the future) that is represented as the emergence of increased liberation from the capriciousness of Nature. It is a negation of the past and a sublimation of the future. Through labour and invention, acting upon the past as its stock of resources, the present becomes instrumental in the perpetuation of that dialectic, which is always represented as offering realisable future rewards. Hence, it is futile to oppose the technological society with a vision of human liberation and individuality because it is already based upon the metaphysical postulation of its own teleology of liberation and progress; hence, it will not be negated by such a critique because it will be able to absorb any such critique and represent it as an affirmation. All such critiques can be represented as impatient demands for the liberation and empowerment of all humanity, and this is exactly the same ideal upon which the technological society is premised. All such critiques can be represented as having their heart in the right place, so to speak, but lacking practicality and realism. Marcuse made the error of presuming that the technological society was premised on irrational instrumentalism or will to power – a headlong and mindless pursuit of innovating new means to carelessly examined ends – but this is only true if we limit our analysis of the technological society to its outward appearance. However, once we reveal the metaphysical foundation of the technological society, we can recognise its moral and ideological essence as an idealistic, dialectical construction of the means to confront Nature and technologically construct "a better future". However, according to Marcuse, human beings are to realise Nature's inherent possibilities in accordance with representations of Nature's potential. This shows the extent that Marcuse's own philosophy of technoscience was very much indebted to the societal gamble and mechanical realism, once reason had been purged of the one-dimensionality of the technological imperative, and instead adopts a dialectical relation between actuality and potential. One finds Marcuse's clearest statement of his faith in the societal gamble in his *An Essay on Liberation*:

Freedom indeed depends largely on technical progress, on the advancement of science. But this fact easily obscures the essential precondition: in order to become vehicles of freedom, science and technology have to change their present direction and goals; they would have to be reconstructed in accord with a new sensibility – the demands of the life instincts. Then one could speak of a technology of liberation, product of the scientific imagination free to project the forms of a human universe without exploitation and toil.[46]

The technological society is a societal aspiration, rather than a concrete reality. Modern society is a complicated and heterogeneous mix of diverse and often contradictory tendencies, aspects, characteristics, parts, dimensions, levels, classes, ideologies, and cultures. The capitalist economic and technological imperatives co-exist within the complex reality of the modern society as societal demands and challenges, which operate upon the direction of the development of modern society. Often these conflict with one another and generate social incoherence in the ongoing transformation of social relations, structures, and institutions, most evident when its innovation and development is directed to overcoming the disasters and problems caused by its prior inventions and industrial processes. The modern representations of human beings as rational agents – understood in economic and technological terms – is a consequence of technological and scientific progress, as theoretical abstractions, which entail ideological commitments. Often these theoretical abstractions merely functioned as an instrument of propaganda or self-justification, while little effort, if any, was actually made to implement them in practice. As Mumford put it,

> Before the paleotechnic period was well underway their images were already tarnished: free competition was curbed from the start by the trade agreements and anti-union collaborations of the very industrialists who shouted most loudly for it.[47]

However, since the collapse of the Soviet Union, industrial capitalism has ceased to be an ideology, directed to order and prioritise the development and trajectories of technological innovation and infrastructural construction, in accordance with a vision of the best of all possible worlds. It has become a mechanised system of maintaining social inequalities that interconnects technologies with the economic systems of production, exchange, and consumption. The interconnected dominance of capital and technique in modern society have pervaded to all areas of the bodily passions, including sex, sport, and war, and when combined with mass media, all these aspects become transformed into spectacles of mass entertainment and commentary. The media domination of mass democracy has simulated public debate by transforming it into the presentation and dissemination of opinions and moral positions. These become off-the-peg objects for adoption and discussion within a system of mass democracy, within which capital, technology, and mass media form a totalitarian machine, and culture and propaganda become one and the same.[48] This totalitarian unity represents the production and consumption of standardised goods, services, and entertainment as being the modern form of the natural order, expressing and reflecting our individuality through choices in modes of conformity. Of course, this representation is one that is designed to naturalise the position of the social elite that owns and controls the access to capital, technology, and mass media, to increase the perception of the artificiality and impossibility of any real social change in the social order. According to such a perception, opposing capitalism would be an irrational act of opposing natural laws, on the basis of an imagined, alternative society that was doomed to fail because it contradicted fundamental human nature and the natural order of things. However, industrial capitalism is no longer based upon a liberal capitalist ideology – in either its anarchist or libertarian forms – directed to improving society by creating the conditions for individual endeavour and prosperity, in accordance with an Enlightenment vision of humanity, but has become an unreflective, unquestioning reproduction of itself – without any coherent vision of the society that it is constructing. Industrial capitalism has become irrational, except as a political and economic means of sustaining the power base of a social elite.

4
The Confrontation with Nature

Human existence is represented as being rationally perfected through labour transforming the natural world into a better artificial one. The technological society and its future innovation is represented as an inevitable consequence of natural evolution and laws, replacing the natural world with itself, while being idealised as the rational consequence of the application of natural science to the construction of society. This entails the fundamental idea that human freedom does not consist in an independence from natural law but, rather, in the knowledge of natural laws and the possibility of making them dialectically work towards definite ends. Supposedly the knowledge of natural laws and mechanisms provides the promise of instrumentally increasing the material possibility of liberating us from the organically evolved alethic modes of our "natural state". This permits a transformation in the concept of labour into the primary relation between human beings and the material world and has profound implications for the social ontology of being and the construction of human experience. The role of a concept of mechanism has not only been essential for the intelligibility of the ontology and epistemology of experimental science and how it has been used to understand Nature, but this concept presupposes instrumentalist conceptions of technology and technical rationality, which have resulted in the dominance of the positivistic interpretation of natural science since Thomas Hobbes wrote and published *Leviathan*.[1] Positivism rests upon a conservative acceptance of a particular set of techniques, without any regard to the contexts and processes through which those techniques were developed in order to satisfy socially emergent goals and projects, while presupposing the metaphysics that allowed artificial processes to be represented as a neutral means to explore Nature by testing theory against experiment. As Horkheimer and Adorno put it, "Such neutrality is more metaphysical than metaphysics."[2] Thus the history of positivistic science is a sequence of metaphysical commitments to particular stages of technological innovation and their necessity for the construction and implementation of particular technologies, as secured through the consensus of scientists and technicians. The positivistic interpretation of science and technology presupposes metaphysical and technological determinism. However, the history of the construction of modern society is the unfolding story of the substantive construction, acceptance, operation, and transformation of the technological processes that brought human beings together and ordered social relations and institutions, as well as the cultural refinement of representations that allowed specific interpretations of the meaning and value of technology and science, in the ongoing universalisation of the Enlightenment vision for humanity. Within modern society, science and technology permeates and mediates all human relations, including our relation with Nature, each other, and our private reflections upon our own nature. To divorce natural science and technology from its history (i.e. to give it a positivistic interpretation) is to both remove it from the social context within which it obtains meaning as a human pursuit, as well as its intelligibility, and to conceal the contingent purposes upon which scientific researches are posited. Once we begin that historical analysis of the meaning and value of technology within our society, then not only are we compelled to recognise its meaning and value are being changed through constant technological innovation and dissemination, but also that the ongoing project of the scientific and technological development of society is an experimental, societal gamble in the creation of an artificial world – a technological society – that promises to be better for human beings than the natural world.

As Horkheimer and Adorno argued, the positivistic degeneration of the Enlightenment into the acquisition of information and techniques has increasingly lead to the decline in philosophical education and critique, while "humanity, instead of entering into a truly human state, is sinking into a new kind of barbarism".[3] The positivistic neglect of historical and philosophical investigation of the social origins, meaning, and purposes of modern science – within the social sciences as well as the natural sciences – has served the prevailing political and economic *status quo* very well. Due to the suppression of social critique of the positivistic technosciences, as being completely in the service of capital, scientists and engineers become isolated as functionaries within the machinery of capitalist economic imperatives and nationalistic propaganda. Even though individual scientists may well act purely out of an intellectual curiosity into "the workings of Nature", while they operate under a conception of rationality and testing, they are bound up with technological innovation, and the truth status of any scientific theory or results is immediately bound to estimations of its practical value and successes within the ongoing development and implementation of techniques and technology because funding for research and the scientific interpretation of its results are intimately bound to commercial, civic, and military ambitions. The epistemological validity of the technosciences is derived from their ontological instrumentality for the construction and further innovation of the technological society. The invariant laws discovered by the technosciences are themselves abstractions underwritten by their ideological function in supporting and satisfying the interests of technological innovation. This is not simply a matter of technoscientists conforming to the vested interests of the owners of the means of production of research and development, but, rather, it is the case that the selection, trajectories, and interpretations of research are made in accordance with an evaluation of the possibilities of future research, as well as being made in accordance with the interests of those who pay the bills. It is not the case that capitalists dictate the outcome of technoscience, but it is the case that the technosciences have been set up from the onset to only consider the outcomes that are instrumental in generating further innovation. The outcomes of research are interpreted in accordance with the contingent cognitive and communicative resources available to the technoscientists to situate these results in the ongoing work of technoscience, as well as satisfying the priorities and needs of those who fund the research. These particularities of these results are not controlled and defined by intentionality and expectations, given that they are products of a complex technological framework within which those intentions and expectations are situated, but they should not be considered to correspond to some natural order either. In fact, the participation of Nature may well be that which both empowers and frustrates human aspirations, but it is impossible for the technoscientists to interpret their agency in correspondence with a natural order, without arbitrarily doing so, unless they interpret their work in terms of the metaphysical precepts of mechanical realism. The whole technoscientific enterprise is itself bound up with the societal gamble interpreted in relation to mechanical realism, as being a natural development of human organisation and labour in accordance with the discovery and utilisation of natural mechanisms and laws in the construction of the technological society. Since the onset of modern science in the sixteenth century, the definition of rational conceptions of science have been premised upon a mechanical realist interpretation of ontological truth being identical with technological power. Henceforth, the philosophy of science became reduced to an epistemological debate between positivists and realists about the possibilities and limitations of the representation of technological power as the consequence of the rational implementation of natural law in material practices: whether causal theories correspond to underlying reality or whether science is limited to empirical descriptions of events. Both philosophies of science conform to the epistemological reduction of knowledge to repeatable and communicable techniques and their associated representations, publicly available to a community of specialists for immediate test in terms of its implementation in ongoing technological innovation.

The art–nature distinction and the societal gamble

Aristotle distinguished between arts that imitate Nature, without changing it, and arts that carry things further and complete Nature, such as medicine, agriculture, gymnastics, and politics, which realise the potential of natural beings in ways that would not naturally occur.[4] However, even though Aristotle considered some arts to lead Nature to a higher state of completion than it would otherwise achieve, he maintained a distinction between art and Nature. This distinction was profoundly influential through the ancient and medieval world. According to Galen, medical art was the servant or agent of Nature (*ars ministra naturae*) that perfected the body.[5] Pliny the Elder stated that perfumes and dyes made by human beings had conquered their natural counterparts in order to improve upon Nature.[6] Pliny described the mechanical arts as doing something artificial by forcing things to act in ways that were contrary to their own inner tendencies, but allowed human beings to conquer Nature. Both ancient and medieval proponents of Aristotelian mechanics maintained this distinction, while affirming the conquest of Nature through using machines against Nature (*para phusein*) to produce "marvellous motions", such as lifting heavy bodies using a pulley.[7] As William Newman argued, the two themes of imitating and perfecting Nature were brought together into the enterprise of alchemy and its goal to emulate Nature, transforming and purifying natural substances, creating new substances, and, ultimately, to replicate the act of the creation of life and make a human better than made by God.[8] Alchemy was an art that not only sought to reproduce natural products in all their qualities (not merely their semblance), but also sought to complete and perfect the form and order of Nature. According to Newman, alchemy became central to an academic debate during the medieval period about the abilities and moral limits of human beings and art in the confrontation with Nature. For some, alchemy represented the ultimate assertion of human power over the natural world, while for others, the assertions of alchemists infringed upon the power of God, aiming to turn human beings into a creator on the same level as the divine. Newman put the point quite nicely,

> The charge of "playing God", commonly levelled against the pioneers of genetic engineering today, was already raised against those medievals who would change the order of the natural world.[9]

As Newman pointed out, the idea of improving natural human generation had been a topic of discussion throughout the medieval and early modern periods. During the sixteenth century, advocates of alchemy (especially the followers of Paracelsus) transformed the ideal of human ingenuity from being that of developing the means to synthesise artificial gold to being that of producing an artificial human being.[10] The utopian speculations of this medieval and Renaissance periods were strikingly similar to nineteenth- and early twentieth-century debates about eugenics and our contemporary debates about the ethical and religious issues surrounding modern genetic engineering and *in vitro* replication of life.

> in the modern world of bioengineering and genetic wizardry, the ever-growing possibility of ectogenesis holds no less a grip on our own visual sensibility, even if its explicit association with alchemy has been lost . . . predicted results of ectogenesis, cloning, the farming of women, and genetic engineering were prefigured by premodern fears that included the production of a diabolical master race, the reduction of women to the status of a hollow incubator, and the prenatal modification of intelligence and gender – all issue that our ancestors found fascinating and at times abhorrent, just as many of us do today. The wellsprings of these dreams or nightmares run deeper than any modern bioethicist or free-market promoter of biotechnology can possibly imagine.[11]

Of course, the exaggerated claims made by the advocates of alchemy were in inverse proportion to what could actually be accomplished, and nobody today suggests that modern science could create anything as fantastic and powerful as the magical golem or homunculus, but the contemporary advocates of biotechnology similarly subject us to fantastic claims and promises, such as the end of world hunger, immunity to all disease, the end of ageing, the prevention of birth deformities, flawless beauty, high intelligence, perfect memory, complete birth control, total cellular regeneration, human immortality, and so on.[12] This shows that our current debates about biotechnology are situated within a long tradition of debating the relation between the artificial and the natural in terms of human arts and the ability to manipulate natural beings for human purposes.

Newman argued that the alchemists of the late thirteenth century and early fourteenth century blurred the traditional Aristotelian distinction between art and Nature because they operated under the principle that things brought into being through art were artificial with regard to their mode of production but natural with regard to their essence.[13] This was a point of departure from the Aristotelian *scientia* and pre-empted Francis Bacon, but, as Newman pointed out, this was grounded in an interpretation of the ambiguity in Aristotle's discussion of roasting and boiling in the *Meteorology*, on whether the art of cooking imitates natural heating; whether there is a meaningful causal difference between the two or whether they are basically the same process.[14] Newman argued that it was this ambiguity that caused considerable debate among medieval scholars about the status of alchemy, whether it was an art that imitated or perfected Nature, and whether its products would be identical with natural substances or mere imitations in appearance. The debate between alchemists about whether there was a distinction between art and Nature shows that the idea of parity between artificial and natural processes (heating, for example) existed prior to the Scientific Revolution, which permitted an intelligible debate about whether it was relevant whether a process occurred naturally or artificially. Furthermore, as Newman argued, by combining this ambiguity with a focus on the process, rather than the product, alchemists represented the products of their art as natural, providing that the artificial processes utilised natural processes. By the late sixteenth century, alchemists such as Bernard Palissy were increasingly working with a blurred distinction between the artificial and the natural.[15] In his *Discours Admirables*, published in 1580, Palissy argued that the art of the potter and the natural generation of minerals were based on the same principles, and he proposed his theory of stalactite, mineral, and fossil formation using ceramic pottery as an explanatory trope. It is only a small conceptual step from the idea that natural and artificial processes share the same *modus operandi* to the idea that experiments based on artificial processes and interventions can be used as a means to understand Nature. Newman argued that, in the seventeenth century, at the height of the Scientific Revolution, Robert Boyle and Isaac Newton, *et al.* were employing the arguments framed by the art–nature debates of the thirteenth- and fourteenth-century alchemists.[16] Hence, Newman argued that, despite all its failings, alchemy was an investigative and experimental science, which was practised for the purpose of investigating the character of Nature by means of operations performed in the laboratory. The case of alchemy not only shows that medieval *scientia* involved experimental investigation into how to artificially manipulate natural processes, but, more importantly, discussions of the limits and powers of alchemy provided the background against which the debate about the possibilities and limitations of the nascent experimental philosophies of the seventeenth century emerged. The alchemists' arguments for laboratory experiment and the replication of natural entities by using artificial processes led to the experimental philosophy of Francis Bacon *et al.* Newman argued that Bacon owed a considerable debt to alchemical literature, and the experimental sciences of medicine and chemistry continued and refined the efforts of medieval alchemists.[17] Bruce Moran also argued that there was not a sharp distinction between alchemy and chemistry in the seventeenth century.[18] He argued that there were many intriguing precursors of modern ideas about the particulate nature of matter, the biochemical paradigm of life and disease, and Newtonian gravitation. Alchemy was not merely the practice of mystical adepts and magicians, but was widely practised by doctors, artisans, and midwives. Even though alchemy may well have been based on theories that were fundamentally flawed or fanciful, Moran argued that alchemy provided the template for investigation into natural processes by basing observation on experimentation and intervention. For example, the alchemical procedure of distillation became the fundamental method of analytical chemistry, and the alchemical goal of transmuting base metals into gold led to the understanding of compounds and elements. According to both Newman and Moran, alchemy was an important and fertile groundwork for the development of the new sciences and the Scientific Revolution.

However, Newman and Moran neglected the profound influence of mechanics on the development of the experimental philosophy of Francis Bacon *et al.*[19] As I argued in Chapter 2, the Renaissance developments of the Aristotelian tradition of the mathematical science of mechanics were crucial for the emergence of the new sciences and the Scientific Revolution. I accept Newman's point that the art–nature debate was initiated and developed by the medieval alchemists (of the Aristotelian tradition), but we need to also address how that debate was also related to the practical arts and the mathematical science of mechanics, from within the medieval debates on the nature of mechanisms. Once we take the emergence of mechanical realism into account, this shows that physics, chemistry, and medicine grew out of a long medieval tradition of debate and experimentation, rather than being radical departures that began during the Renaissance, which completely dissolved the distinction between art and Nature. The fundamental conceptual difference between Aristotle's physics and that of Bacon, Galileo, and Descartes was that, according to the former, machines lacked an internal principle of change, and according to the latter, machines, after they had been brought together and constructed, operated according to natural law. Mechanical realism had become established as a self-evident truth that permitted art to be used as a means to explore underlying natural principles. Hence, during the seventeenth century, it was widely considered, by mechanical realists such as Isaac Newton, Robert Boyle, and Robert Hooke, to be the case that it did not matter whether a substance was brought into being by artificial or natural means. If artificial and natural gold shared the same properties and qualities, withstanding the same tests, then both would be identical. The natural and artificial metal would only differ in their mode of production, not in their essence. However, the idea that there was symmetry between the natural and artificial did not stop at the contemplation of the ontological equivalence of artificial and natural substances. Modern science and technology grew out of the Aristotelian arts of mechanics, alchemy, and medicine, and was directed to improving the human condition by using scientific knowledge to perfect the world. The practical value of the new sciences was central to the whole enterprise of the development of the new sciences from their very beginnings. Since Francis Bacon proposed that humanity could better its conditions and be liberated from the dictates of its organic state by using the manual arts – the happy match between the human mind and the nature of things – to provide inventions (such as gunpowder, the compass, and the printing press) that would establish humanity as the master of Nature, the Enlightenment project has been bound up with the societal gamble (underwritten by the scientific world-picture) in the possibility of constructing a world within which human beings could construct a better world, within which scientific knowledge would liberate humanity from the capriciousness of Nature and the pernicious influence of superstition. For Bacon, the new sciences promised such things as the prolongation of life, the mitigation of pain, the retardation of age, and the restitution of youth. The new sciences were to create a utopia, within which the scientists, as the denizens of Solomon's House in *The New Atlantis*, among other things, were to direct their efforts to provide human beings with new powers, making new experiences possible, and bringing everything within our reach.[20] *The Book of Genesis* story of Adam and Eve's fall from grace, being hurled from the Garden of Eden into an uncultivated wilderness of suffering and struggle, which required labour and the "sweat of their brow" in order to survive, was central to Francis Bacon's vision regarding the goodness of a society based on practical knowledge and arts. The idea of the human manifest destiny to use science and technology to build a better world, a new paradise on Earth, was that of a second creation based upon our God-given reason and intelligence in order to discover all the natural means that God had place in this world for us to use to our advantage.[21] Bacon's writings are a statement of the societal gamble in the benefit of technology and science for the human condition, within which our natural state is one of confrontation with Nature. The societal gamble is that science would ameliorate scarcity and universally satisfy all human needs. Within the scientific world-picture, human beings and the modifications of the natural world that human beings impose upon the pre-existing order of things are taken to be a part of Nature themselves because human beings and our capacities to change our environment are taken to be natural. Through this interpretation of how the practical activity of human beings is possible, Nature is represented as being implicated in its own modification. Without the metaphysical precepts of mechanical realism, this interpretation would be completely arbitrary and poetical, and, hence, the rationality of the scientific world-picture and the whole project of developing the experimental sciences as natural sciences entail the presupposition of those precepts. Thus, even though the ongoing development of the technological society places human beings at a far remove from any natural and organic state of animality, it is only the artifice which is implemented and developed in accordance with natural laws that is considered to have a chance of working and, hence, being rational and progressive. The representation of Nature as an independent and objective reality exists as a result of the metaphysical unification and transformation of collective human experience of the technical limitations, possibilities, and failures, realised through interventional material practices.

As I argued in Chapter 2, the development of craft practices and the innovation of instruments were central to the work of Galileo, Descartes, and Newton, and we can see a clear technical intention in the development of natural philosophy during the seventeenth and eighteenth centuries' drive for the achievement of commercial, political, and military advantage in the competitive contexts of European ambitions. From its onset, modern science was premised upon the manipulation of Nature for the empowerment and benefit of human beings. Once mechanical realism was presupposed, the task of acquiring and testing knowledge became one of refining the scientific world-picture by articulating a system of mechanisms and resources. This shows how modern science attempts to understand the nature of matter by representing material practices in technical forms that identify a set of machine performances. We can cognate, manipulate, and control these machine performances in terms and representations that we know because we have created them for ourselves in order to help us organise material practices better, as something objectively before us, graspable through bounded technical rationality as a system, developed in accordance with the practical problems of ongoing scientific research, as well as the problems of wider society. This promises us the certainty and power of knowing and manipulating the world in terms of representations that we have created for ourselves on the basis that they have been instrumentally tested in material practices. The seventeenth-century scientific world-picture was refined in order to explain the technological development of machines in terms of increasingly fundamental mechanisms. Technology was represented as the consequence and mediation of all relations between "Man" and "Nature", as being the rational utilisation of natural mechanisms in the ongoing development of material practices, and, hence, it is understood as the application of the causal powers discovered by the experimental and practical sciences. The acquisition and control of new technological powers was equated with the increased understanding of reality. The awakening of "Man" became represented within the Enlightenment project as being the recognition of the truth that a rational understanding of technological power provides a rational understanding to all relations and changes. The Enlightenment project was much more than a struggle against the authority of Church, the divine right of kings, and superstition. Of course, it was a process of establishing a means of rationally using a scientific world-picture as a standard by which such things could be shown to be arbitrary, but it also involved a fundamental faith in the progressive nature of scientific knowledge as being something that would liberate human beings from ignorance, squalor, fear, and vulnerability. Nature became represented as the objective source of all power, the neutral arbiter of truth, and something that must be overcome. Hence, the development and refinement of the scientific world-picture alongside the innovation of methods, techniques, and instruments became the basis for the modern conception of rationality in terms of bounded technical rationality. The focus upon material practices and the satisfaction of individual interests through technology preceded the nineteenth-century Industrial Revolution. This individualism was itself a consequence of the beliefs in the primacy of the techniques of rational and reasoned discourse and the unitary material relation between "Man" and "Nature" as disclosed by the sciences. These conceptions were a consequence of the societal gamble rather than its conditions and were the conditions for the Industrial Revolution and the mass participation in the societal gamble to be represented as progress, social evolution, and human destiny. The direction of scientific research was immersed and stimulated by the practical problems of the society for which the sciences were seen as both beneficial and progressive, and as new machines were invented, the refinement of the scientific world-picture was extended to encompass more of reality. By the nineteenth century, the mechanical realist dream of explaining the totality of existence in terms of underlying mechanisms had become realised as a positivistic, technical task, bound together with technological innovation and the satisfaction of the ambitions of society. Once the nature of human being was defined and represented in terms of technically rational agency, then the human intervention in the world becomes represented as the means through which human beings can come to discover human nature in terms of our capacities and limitations. It is quite ironic to note that, within the scientific world-picture, the greater the degree of technological intervention and mediation is represented as an increasing convergence between human awareness and objective reality: the greater degree of artificiality is taken to be a step closer to harmony between "Man" and "Nature".[22]

Nature and the technological society

In *The Question Concerning Technology*, Heidegger attempted to relate the essence of technology to truth because he was concerned with preparing a way in which we could develop a free relationship with it.[23] He criticised both the instrumentalist and anthropological definitions of modern technology because, even though he accepted that both were superficially correct, they have made us blind to the essence of technology and place us in an unthinking relation with it. The anthropological definition is that technology is a human activity and the instrumentalist definition is that technology is a means to an end. While Heidegger agreed that technology is a human activity in the sense of positing ends and procuring the means to them, and he accepted that it is also an instrument in the sense that equipment, tools, and machines are used to make things, he argued that we need to question the meaning of "the instrument" and ask to what do means and ends belong, if we are to understand the essence of technology. Heidegger described modern technology as a mode of disclosure that challenged a tract of land to yield coal and ore, disclosing the earth as a coal-mining district and the soil as a mineral deposit, while air is set upon to yield nitrogen for use by industrial agriculture. Heidegger used the example of the hydroelectric plant to describe the way that the Rhine is disclosed as hydraulic pressure for an interlocking complex of turbines, electromagnets, power stations, and a network of cables, set up to provide electricity. The river is damned up into the power plant and is transformed into a power supply. Even to the extent that it is still a river in the landscape, it only remains so as an object for the tourist industry. Heidegger lamented that we are rapidly approaching a time when there will no longer be any natural world at all. Nature is disclosed as a source of energy that can be extracted and stored for future use. Heidegger termed this availability for future use as "standing-reserve". The essence of modern technology is an imperative – termed by Heidegger as "Destining" (*Geschick*) – that challenges human beings to exploit Nature by disclosing it as standing-reserve. Once human beings respond to this challenge, then we are set upon by modern technology, gathered together and ordered into a mode of disclosure of everything into standing-reserve, which Heidegger termed as "Enframing" (*Ge-stell*). Destining does not involve any technological determinism, in the sense of some inevitable or fated path of technological development, because its direction is dependent upon human participation in the way of disclosure, even if human beings do not control what Enframing discloses. Heidegger argued that the essence of modern technology – as Enframing – is not something technological (in the same way that pistons, rods, and chassis are technological), and, consequently, the anthropological and instrumental definitions of technology conceal it by treating it as if it were. This threatens our free essence if we surrender to it.

For Heidegger, the mathematical was central to the definition of modern science because only the quantifiable aspects of reality are available to scientific research, and, hence, the scientific use of technique is that of reducing the possibilities of investigation to the calculation of numbers.[24] According to Heidegger, the Destining of human beings to take up the challenge of Enframing, disclosing Nature as a standing-reserve of energy, was first displayed in the rise of physics as an exact science, which represents Nature in terms of forces that can be calculated in advance, and orders experiments in order to determine how Nature responds when it is set up in this way. Heidegger argued that modern physics was "the herald of Enframing" that prepared the way for the essence of modern technology, by disclosing Nature as an orderable system of information that is calculable in advance, and the Industrial Revolution employment of exact physical science. Heidegger was critical of the way that *techne* is treated as the ultimate virtue in the modern age.[25] For Heidegger, the ancient handicrafts were a different mode of disclosure from modern technology because they participated in "bringing-forth" (*poiesis*) beings into the world as ends-in-themselves. They were intimately bound up with *aletheia* (truth) as a mode of disclosure and presencing of the real. This truth was bound up with modes of completion and perfection and did not correlate with the definition of truth as "correctness" (*veritas*).[26] It was implicit to Heidegger's understanding of modern technology that it mediates our experience of Nature by disclosing those aspects of Nature that are amenable to being disclosed in this way.[27] However, as I argued in *On the Metaphysics of Experimental Physics*, experimental sciences are on the boundary between craft and technology – on the boundary between *Ge-stell* and *techne* – thus the disclosure of instrumentality became represented as the discovery of truth, as *aletheia*, through providing causal theories of how *poiesis* was possible, modelled in terms of natural mechanisms.[28] It is only after the interventional processes have been removed from the account are these causal theories are presented as a set of linguistic propositions, as *veritas*, which are available for empirical test and logical analysis. However, once we recognise that experimental science can be understood as a technological process that is presented as a natural science on the basis of hidden metaphysical precepts, used to interpret the contours of human intervention and machine performativity, then we can see how scientific theory is related to technological practice to produce new machines and novel phenomena, and how knowledge is produced through experimentation and related to natural phenomena. This entails metaphysical presuppositions about the form of reality, which underwrites epistemological assumptions about scientific rationality, knowledge, and progress. It is this metaphysical interpretation of the meaning of human efforts to disclose truth in terms of technological power that permits experimental physics to participate in discovery. For Heidegger, the essence of materialism was concealed by modern science not only because it asserts that the world is solely comprised of the physical interactions between particles of inanimate matter, but also because it presupposes a metaphysics that reduces every being to material for labour. In the 1949 version of his *Letter on Humanism*, he noted that modern science had become the new metaphysics into which philosophy was becoming dissolved.[29] The unity of this metaphysics was unfolding, in a new way, in the science of cybernetics.

Norbert Weiner in his famous book *Cybernetics: Or the Control and Communication of the Animal and the Machine* (1948) signalled the most positivistic stage of technoscience within which the scientific invest-igation of all physical processes could be reduced to feedback systems of information and control. The subsequent extension of cybernetics into the biological and human sciences is the positivistic culmination and unification of the technosciences. Cybernetics theory reduced the stratified ontology of realism into an ontological dialectic between inform-ation and background in such a way as to disconnect the concept of mechanism from causality and, instead, represent it in terms of a concept of control, defined in terms of non-linear feedback loops between signals and responses. Henceforth, the cybernetic system became the dominant explanatory trope in all technosciences, ranging from neuro-logy and psychology to genetics and James Lovelock's Gaia hypothesis. This has become the latest refinement of the scientific world-picture within which all phenomena are described in terms of dynamic, non-linear feedback loops between signals and responses. Nature has become a non-linear sum of all signals against a background of universal noise, a total resonance peak of order against a background of randomness and chaos. Henceforth, technology and nature processes are unified under the same category of being, represented, understood, and interacted with, as a system, and the human relationship with natural processes becomes inherently one of management, information, and control. In many respects, the spontaneity of Nature has been reduced by the tech-nological society into its capriciousness, those aspects that escape the enframing grasp of the technological society are taken as pernicious, undesirable, and chaotic. Moreover, the technological society removes natural beings and replaces them with artificial ones to limit Nature's power to spontaneously or periodically cause havoc, failure, or decay in the technological society. The societal gamble simultaneously postulates the goodness of the technological society, whilst representing that arti-ficial world as being a natural consequence of human rationality and evolution, gathered together and ordered by technology, set upon the material practices in which the techniques and machine performances are ontologically connected into a system that is unified with the natural process, while replacing natural beings with artefacts. It achieves its own trajectory – its invariant identity and reality – through the whole process of trying to stabilise the coherence of the system during the processes of integrating all aspects of the human lifeworld within the system. The substitution of the technological society in the place of the natural world underwrites the phenomenological experience of Nature as Otherness – beyond human experience and cognition – as something that needs to be confronted, pacified, and replaced with something that can be controlled, manipulated, and predicted.

The transformative power of each machine within the technological society can only be understood in terms of the non-linear relations within the total system within which it is situated. Thus, given that every technology is continually transformed and refined during its implementation, it does not have any power in isolation. The aesthetic judgement of the success of integration of any technology in the ongoing perfection of the system is one that is a judgement upon how it functions within an intelligible analysis of the ongoing systematisation of all labour. This judgement is premised upon a reduction of quality to reproducible functional efficiency within an ongoing process of technological innovation. Within this process, the scientist is transformed into an agent for whom the aesthetic notions of technical excellence and ability are themselves abstracted from a perpetually ongoing process of transformation and refinement. Thus the agency of human beings *qua* scientists becomes appropriated in terms of the systematisation of all labour, transformed via technique into a set of procedures and practices, whilst the abstraction of technical excellence and ability permit the labour process to be one in which the human being is the agent and the machine is the patient in which the irreversible and unpredictable nature of technological innovation is represented as being the product of human creativity and control. However, in reality, the human creativity and control is itself a product of the whole system. Human beings *qua* individuals have no productive abilities whatsoever; it is to the extent that human beings *qua* individuals are themselves coherently integrated as agents that they acquire the power to produce and create anything at all. Technical rationality is bounded by the system and is an emergent property of the system, rather than being imposed upon the system from without. The process of production and reproduction transforms the system, as its emergent properties are transformed through the whole process of creating and reproducing coherence within the system – hence, productive agency is non-linear; it is both produced by and produces the process in which it is integrated as a component. The project of constructing the technological society is a societal endeavour to create a self-referential system – a complex *autopoiesis*, which generates and sustains itself – as a substitution for the natural world. Technoscience is directed within the project of the autopoietic creation of the technological society, as the final goal, in order to allow human beings absolute independence from anything outside technological control and manipulation. Thus the system becomes autonomous and self-referential, acting as a kind of autopoietic parasite upon Nature.[30]

Once the technological society is understood as an autopoietic system, created from the material efforts to appropriate natural beings in order to create an artificial substitution for the natural world, then the objectivity of the natural laws discovered by the technosciences can be understood as emergent properties of the autopoietic system, rather than correspondent with immutable laws that transcend the system. These laws are the techneic idealisations of our alethic modalities, which are periodically refined and transformed as the system achieves further levels of complexity and is capable to innovating new kinds of machines, techniques, and rules of cognitive representation. The economic and social relations emergent from within the ongoing construction of this system are indeed based upon the dialectical relation between our material conditions and the exploitation of technological powers, but this relation does not need any conscious connection with any natural substratum of reality in order to empower those relations. Their power is also an emergent property of the system. Technoscience has become an autonomous and dominant mode of creativity that dialectically orders and transforms the system. It increasingly mediates all human relations, experiences, expectations, and intentions, creating and transforming modes and systems of life, to such an extent that these become increasingly artificial and complex properties of the system, which only have an increasingly abstract and symbolical relationship with organic, natural being. In short, it constructs and transforms the ontological basis of human action, which means that it dialectically constructs and transforms reality, while representations of that process are products of the process. The positivistic representation of the natural sciences (as being a method of obtaining the neutral facts by means of logic, observation, and experimentation) is only possible because it presupposes the fundamental representations necessary to make observations and experiments in the first instance, as well as the fact that science is a social process situated within and emergent from economic and political realities of industrialisation and mechanisation. The idea that the construction of machinery, in itself a historical outcome of an evolutionary process, performed in accordance with natural laws in order to satisfy biological necessities, is itself an emergent product of the system that is metaphysically understood. The assumption that the system is only possible because of objective material conditions and natural laws is an assumption that is not an empirical proposition because there is no way that this assumption can be verified or falsified without presupposing metaphysical precepts in order to construct a test for this proposition.

The positivistic interpretation of the natural sciences and technology has been heavily criticised by Marxists, for whom natural science is a social phenomenon and therefore can only be understood in terms of a dialectics between theory and practice, understood through productive relations, but they accept the ontological foundation of this dialectics in the material reality of Nature, given that truth is tested through its practical use-value.[31] Marxists (like all realists) argue that we should comprehend concepts as reflections of real things (i.e. material objects and structures) rather than real things as reflections of concepts.[32] As Leszek Kolakowski put it,

> The categories into which this world has been divided are not the result of convention or a conscious social agreement; instead they are created by a spontaneous endeavour to conquer the opposition of things. It is this effort to subdue the chaos of reality that defines not only the history of mankind, but also the history of nature as an object of human needs – and we are capable of comprehending it only in this form.[33]

But this still leaves the questions of how reality was comprehended and whether and how the process of comprehension transformed that reality. Even though it may well seem realistic to assert that reality is the criterion for the correctness of thought, this fails to recognise the extent that reality is a total process and thus the participation of thought is not only part of its totality, but is an essential aspect of the representation of reality as a totality and how its parts relate to that totality. Hence, while Marxists are critical of the social sciences, they adopt a realist acceptance of the scientific world-picture and accept a somewhat naïve correspondence theory of scientific truth, especially when they argue that capitalism perverts scientific truth and technology. Georg Lukács was critical of the orthodox Marxist analysis of technology as an autonomous means of production that inevitably improves production and efficiency, being a necessary force in human emancipation.[34] He argued that the orthodox Marxist analysis involved a deterministic faith in science as being a liberating force, which by transforming the forces of production will inevitably transform the social relations of production and social reality.[35] Lukács agreed that capitalism emerged as a result of class conflict and is dependent upon the technologies of industrial production, but argued that science and technology are social products that cannot be separated from the emergent relations of society upon which they depend. Science and technology are no more liberating or progressive forces than any other aspect of society. Social relations of production impact the nature, direction, and rate of technological development, reflecting the interests of dominant groups embedded within the social structures and institutions that shape the directions of the sciences. Consequently, understanding scientific and technological change must be centred in an understanding of power relationships. Science and technology reproduce and fortify the social hierarchy, and those that are able to associate themselves with technical or scientific "expertise" are able to command special status within a society that reifies technology as reflecting the logic and structure of natural law. Science and technology become effective methods of social control – directly and indirectly – within a total system of reproducing inequality and privilege. Once we properly situate the objective laws as emerging properties of a social and technological complex, then we can see that the technological society is ontologically based on socially emergent autopoiesis, rather than the rational application of natural laws, which creates its own reality. Rather than remaining in a dialectical relation with our objective, organic conditions for existence, modern industrial society becomes an increasingly artificial society that not only creates its own means of production, but also creates the understanding of its own origin and conditions for its existence, while perpetually transforming its form and content. It only has a contingent symbolical relationship with the natural world, while it continues to appropriate and transform the natural world in accordance with the purposes of its own development.

For Lukács, even though the possibility of social reality is founded upon the material reality of Nature, whatever is taken to be natural at any stage of historical development of philosophy, science, or art are all socially conditioned and mediated in terms of highly complex techniques and abstractions. Whenever "Nature" is related to human beings and our involvement with material reality, then ersatz and increasingly sophisticated human relations mediate the evaluations of the truth, form, content, range, and objectivity of these representations of "Nature". Modern science, art, religion, and philosophy are all premised upon mediating representations of Nature, given in terms that are constructed and understood on the basis of human material involvements with that which surrounds us and within which we find ourselves.[36] Art is only able to represent Nature as something wholly above history and society because it rests on the ability to represent human involvement and positioning within the world as being a dialogue between "Man" and "Nature". It is only able to do this because it is socially represented as being a mediator between "Man" and "Nature". However, as Lukács argued, the method of dialectics applies to the interaction between subject and object, theory and practice, historical changes to the categories and structures of thought, and so on, and these features are absent from the Marxist understanding of the scientific knowledge of Nature.[37] The scientific naturalism presupposed in Friedrich Engels' interpretation of science demonstrates the mechanical realist basis of the ontology of Marxism from the onset – that social reality was based on the ontological foundation of the natural and experimental sciences in material structures and mechanisms. Engels described the human hand as both the organ and the product of labour.[38] The human hand has required both an evolution of its biological form and a history of adaptation to increasingly complicated tasks and operations in order to become capable of producing intricate works of art. The biological development of the hand was a precondition for making tools, but it is only in the ongoing efforts towards the perfection of the hand through labour that human beings can be said to be achieving mastery over Nature, widening the horizon of possibilities, making new discoveries of the natural properties of materials, and developing the sense of touch. Labour brought individuals closer together through the necessity and advantage of cooperative effort to the individual and was central to the development of language and society. Labour was the foundation of the emancipation of human beings from the harsh conditions of the natural world and with each generation labour becomes more differentiated and improved. The structure of trade and industry finds its model in the family, which is the natural and primary form of human organisation, which became transformed when labour in the family became specialised and performed by hands other than one's own. After this division of labour, the value of the hand retreated into the background of human labour and the mind that plans labour ascended into the position of dominance. Henceforth, all the products of labour and human society were represented as being the products of the mind, and the origin of all human action and intentionality was taken to be thought rather than need. According to Engels, in order to emancipate ourselves from our material conditions, we need to obtain and take advantage of our knowledge of the natural laws by correctly applying that knowledge. Engels' theory of natural science is based on a material dialectics between historically contingent human actions and the objective structures and underlying mechanisms of the natural world. Our increased scientific knowledge implies that we increasingly have the ability to learn and understand the consequences of our actions on the natural world and how these will, in the future, impact upon us. Such knowledge will increasingly reveal the interconnectedness between all things and the human unity with Nature. As a consequence of this revelation, the traditional distinction between mind and body will become increasingly antiquated, until it is finally dissolved. However, we are only beginning to understand some of the natural consequences of human labour and production, and the social consequences of these actions are even harder to predict. To use Engels' examples, the potato famine in Ireland of 1847, which lead to the death and forced migration of millions of people, was not a foreseeable consequence of the spread of the potato, just as the role of alcoholism in the cultural disintegration of North American Indians was not a foreseeable consequence of the distillation of alcohol by the Arabs many centuries before. Columbus could not know that his discovery would lead to the enslavement of Africans, civil war, the struggle for civil rights, and the racism of modern America. The inventors and pioneers of the steam engine could not predict the massive changes that occurred because of the widespread use of this machine. The complexities of social interactions are a consequence of unforeseen discoveries that prohibit such predictions because they cannot be understood as directly following a single set of laws or principles, or even a complex interplay of laws, because they are based on a series of decisions between alternatives, but, due to the underlying objective reality and forces of Nature, these decisions will inevitably lead to unforeseen consequences and problems that require further decisions and adaptations. So, even though the careful scientific collection and analysis of historical cases does offer us the possibility of predicting and controlling some of the consequences of future actions, it does not do so on the basis of scientific knowledge alone. It requires a transformation of contemporary social order and modes of production so as to extend labour beyond the mere satisfaction of immediate wants (without any concern for other possible alternatives and their consequences) in the interest of a ruling class (which in capitalist societies is defined solely in terms of profit and the acquisition of wealth) in order to place our decision-making process and adaptability in relation to a long-term view on the basis of scientific knowledge. However, capitalism resists this scientific development and evolution because it is only concerned with immediate effects of production and exchange, and when capitalists limit the consequence of production and supply to be that of the immediate acquisition of profit, then only the most immediate results can be taken into account within positivistic economic analysis. Capitalist modes of production are only concerned with the first tangible success of the production of commodities and goods. The social and environmental consequences and value of any product is of little or no concern except in terms of its relation to the analysis of supply, demand, costs, and profit.

It is this ontological and epistemological foundation of Marxist theory in the natural sciences that reveals the extent, despite Engel's opposition of the concepts produced by dialectical materialism to Cartesian metaphysics, that Marxist theory presupposes the metaphysics of mechanical realism and the goodness of the technological society. Even though the concepts developed through dialectical materialism are premised on human interactions – rather than rigid categories and abstract definitions – in order to show the transitive and concrete essence of social reality, as the basis of a method directed towards changing that social reality, the whole methodology of Marxist theory is itself founded upon metaphysical ideals regarding the teleology of modern society and the emancipation of human relations. That is not to say that Marxist theory is necessarily premised on some metaphysical determinism or fatalism – as it is central to Marxist theory that all change requires conscious human action and choice – but it does mean that it is based on idealised representations of the form and content of material and productive relations, given in terms of moral and technological imperatives, and, thus, it is prescriptive of what constitutes conscious and rational human action in relation to our material conditions. Technoscience has transformed not only our material conditions, but also human agency, which means that, in part, it has transformed what we are and how we understand ourselves. The objective laws of Nature have always been a reified abstraction of the relations between human beings. The individual human being is situated within a technological framework that is concealed by being represented as the rational implementation of natural mechanisms, which allows the structures and institutions of society to be represented as the rational implementation of knowledge, rather that the consequences of deliberate human decisions.

The construction and development of the technological society is a process of constructing and developing human relations in accordance with the contingent directives and constraints of the technological and capitalist economic imperatives. Lukács described classical economics as being based upon a reification of specific, historically contingent human relations, as if these were autonomous natural laws of exchange, supply, and demand, and, consequently, capitalist economic theory was founded upon a "phantom objectivity".[39] Classical economics, with its systems of laws described in terms of related differential equations, is modelled on modern physics, but the directions of research in modern physics are dependent on the commercial, civic, and military ambitions of those who pay for the research. The point at which economics has some degree of predictive success is exactly the same as those of physics: when its ontology is that of sets of machine performances and their bounded technical rational application as an abstraction within ongoing practices – that is, modern technology and the development of the productive relations. The economic structure of society is dependent upon its technological possibilities, and, hence, the "phantom objectivity" of classical economics is also premised upon the same scientific world-picture as the experimental sciences, but the content of the scientific world-picture is contingent upon the development of the technological framework from which it is emergent and refined. While the positivistic interpretation is based on a reactionary reification of contemporary productive relations and their interpretation in terms of the current content of the scientific world-picture – a fetishism of techniques and technological objects – the Marxist faith in the possibility to liberate science from the intervention of the short-term and conflicting interests of capitalists depends on a faith in the essence of science being directed in accordance with a will to truth. The Marxist faith in the goodness of the technological society within which science would be the driving force of industry and increased productivity and efficiency is itself based on a metaphysical and ideological interpretation of the essence of science as being a good. Within both positivistic and Marxist interpretations, science, technology, and industry are interdependent, mutually enforcing, and transformative; therefore, human self-knowledge is equated with the alethic knowledge of agency and, thus, human being is represented in technological terms.

Positivism permits the act of producing an object as being represented as the immediately correct realisation of the theory associated with that act. Positivism rejects this association as being represented as a causal account (instead focussing upon descriptive isomorphism between the structures of theory and experience), and, hence, removes from the remit of science every question about the underlying reality that makes experience possible. Positivism limits the object of all experience and enquiry to be the outcomes of applying particular techniques and, hence, bases the validity of all scientific facts and theories to their utility for technology. However, given that technology is bound up with purposes, positivistic sciences are ontologically founded on a teleological system that is operational in the organisation and construction of an experimental technological system. The economic totality of the capitalist economic system is ontologically founded on the concrete reality of the technological society, within which industrial production is teleological in both the sense that it is the synthesis of the functions and purposes of distinct technological systems and in the sense that it has a teleological function with the construction of society as a totality. Once we examine the teleology of each and every technological system – in terms of the wider purposes and functions associated with those systems – and relate those teleological positings to the technological society as a totality – in terms of the extent that they are posited as satisfying the technological imperatives of maximising efficiency and productivity – then we can examine the constructive role that those systems have in the form and content of an incomplete and experimental society. Of course, when such a determination can only be performed in hindsight, it is often obsolete even before it is even articulated. This reveals the extent that the technological society is teleologically challenged, perpetually driven, to perform its own innovation in order to satisfy its insatiable technological imperatives, as an unreachable ideal that has form without content. *Techne* is the ideal knowledge that is imagined on the asymptote of the construction of the technological society, but, given that the ontology of the technological society is comprised of the productive and creative relations between human beings, *techne* becomes the knowledge of the rational and universal completion and perfection of society. On a Marxist interpretation, it is essential for the socialist culmination of the technological society – as a liberated, scientific society – that its tacit knowledge and productive power is distributed throughout the whole society. Dialectical materialism inherently represents the totality of history as the process by which human evolution is mediated within the development of the technological society. Thus the demise of capitalism and the dissolution of the proletariat as a class is simultaneously the integration of everything into a totality within which each part is coherently ordered and responsive to the whole in accordance with the technological imperatives to maximise efficiency and productivity. The dialectical method of relating every fact of social life as aspects of the totality of the same historical process, as revealing the essence of social reality through revealing the class struggle between capitalists and the proletariat, has written modern history and all social relations in terms of the ongoing development of the emergent consciousness of the form and content of the technological society as being identical with the emergent class-consciousness of the proletariat. However, once we take Lukács' criticisms of Marxism into account, then we can see how the process of technological innovation dialectically transforms the agency and alethic modalities of human relations, and, given that technology is central to the progression of the science, how it also transforms the scientific categories under which those relations are understood. The integration of all knowledge and action into a perpetually refined and perfected scientific world-picture, emergent and transformed during the whole societal project of constructing the technological society, is a process of the abstraction and representation of a societal totality as if it were a consequence of human evolution in accordance with natural laws. However, this act of integration is one that conceals the extent that both the form and the content of the technological society, at each and every step in its construction, are dependent on historically contingent human choices and, therefore, remain artificial, revisable, and experimental. Once it was metaphysically assumed that human beings can only know something to the extent that it can be made, manipulated, and theoretically represented through material practice, then the form and content of scientific knowledge became based upon a hierarchy of technical interventions and estimations of alethic modalities, represented in terms of abstract natural laws and mechanisms, that presupposed the causal unity behind both technology and the natural world. Human beings became represented as the intersection between these two aspects of the same as being the rational and technical agent through which the latter is transformed into the former. Human beings are represented as being the rational efficient cause that transforms the natural world into the technological society. However, given that the technological society operates on a twofold process of transformation through technological innovation because it dialectically transforms itself using technological objects as both a resource and a means, the technological society is an autopoietic substitution of itself for the natural world – the societal gamble is that this world will be better than the natural one for human beings. Hence, autonomy and universality are synonymous expression of the totalitarianism of the technological society as a substitution for the natural world.

The confrontation with nature and the pacification of existence

There is a long history of the domestication of plants and animals, which in many important respects are artificial hybrids, but in the sixteenth century plants and animals became represented as machines. Even though genetic hybridisation and transgenic organisms were not created until the twentieth century, we can see how the naturalness of these artificial processes finds their roots in the sixteenth-century representation of living beings as being something technological. The extension of technoscience into greenhouses and the creation of an artificial agricultural and horticultural environment is the scientific determination of the exact conditions of soil, temperature, and moisture that are needed for specific plants to grow anywhere, as universal resources to satisfy specific purposes and needs. Through the concept of the *ecosystem*, the use of genetic modification and biological controls is represented as an artificial development of the natural state of affairs, treating ecology like a complex of useful mechanisms open to manipulation and modification in order to liberate agriculture from pests and parasites, within which the criteria of risk, control, and reproducible yield become the only rational factors for deliberation. The transformation of Nature into an ecosystem or biosphere is one that is constructed through the juxtaposition and abstraction of machine performances into a framework of inter-related mechanisms, functions, and models projected over the natural world. Within ecology all Nature is reduced to a series of interacting mechanisms and is a "reflection" of the technological framework from which the science of ecology emerged. Ecology is a product of the technosciences such as physics, chemistry, and genetics, which eclipsed its origin in natural history, and is itself a hybrid and complex technoscience that is formed by creating an intersection between these sciences and projects the values of the technological society over Nature in order to identify and explain the natural world in terms of efficiency, instrumentality, and function. Hence the natural sciences are premised on an identity between truth and value, objectifying behavioural norms as if they were selected in accordance with an objective understanding of their benefit for the organism. Today this is becoming increasingly evident in how modern biological science is attempting to understand what human beings are by changing human embryos and the conditions of gestation. The science of genetics employs the same techniques and representations that it applies to non-human life to such an extent that animals, plants, bacteria, viruses, and human beings are all ontologically equivalent and at the disposal of this new science to test itself by transforming its objects under controlled artificial conditions in accordance with ideas of how those objects can be improved. The scope of biological science has itself been transformed from something at the disposal of human beings to improve the conditions of human life to something that is capable of improving human beings before birth. As John Passmore observed, the ongoing drive to perfect human beings through selection might well result in the evolution of human beings into a new species.[40] Increasingly the idea of using some kind of gene therapy to mitigate disease has not only become widely accepted as a benefit, but there is even a growing sense of a moral obligation for the innovation and use of such techniques. Within the technological society, the techniques of gene therapy are as acceptable as those of dentistry and the only rational basis of deliberation is regarding its efficiency and risks. Thus the question of progress and changing the environment to the benefit of human beings has become reduced to the question of the extent that the environment, including other human beings, is at the disposal and under the control of human beings. The sense of primordial belongingness of human beings to Nature has been fundamentally transformed into instrumentalist conception of the natural world as itself being the manifestation of a natural relation. All is production and selection. Environmentalists are henceforth compelled to argue that rain forests should be preserved because they are ecologically functional in reducing carbon dioxide and providing the atmosphere with oxygen. They are to be valued because of their instrumentality in preserving our existence. Thus the rain forests of Brazil, for example, are represented in technological terms in order to give them instrumental value within a political struggle (mediated by the institutional mechanisms and media of lobbying and policy making) between capitalists and environmentalists that is ultimately a debate as to whether rain forests have more value as timber or as a mechanism to transform carbon dioxide into oxygen (or provide rare herbs for the pharmaceutical industry, much needed tourist dollars and jobs for local communities, etc.). All the other possible conceptions of the value of a rain forest (which are more often than not expressions of awe and respect) are considered to be irrational or subjective.

The objects produced by technoscience cannot be understood without relating them to the goal of the construction of the technological society. Science is an activity within the world. The scientific understanding of the world was set up from the onset of sciences such as physics, chemistry, medicine, and so on, to order and control things in the world with the explicit intention of making the world a better place by providing novel machines, materials, and medicines, and so on. Nature is represented as flawed, imperfect, and in need of improvement to the extent that the essence of human nature is represented as being a natural drive to improve Nature in accordance with human purposes. The practice of science is teleologically postulated as providing human beings with the means of improving the natural world and human intelligence. Of course, human beings have changed and organised the natural world through such things as deforestation and agriculture for thousands of years, as well as controlling, domesticating, and selectively breeding animals and plants, but within the modern technological society life itself has been reduced to a set of basic chemical elements and biochemical mechanisms, and thus it is merely considered a natural extension of farming and husbandry to use genetic science to change the structure of organic life itself in accordance with human purposes. In fact the scientific understanding of all natural phenomena require them to be changed in order to be able to understand them and test that understanding. Adorno and Horkheimer argued that instrumentality is a form of domination that controls objects, violating their integrity and distorting their nature. However, as I have argued above, a closer examination of the technosciences shows that they are premised upon the representation of the world in terms of its instrumentality, but this is not so much an act of domination, rather it is an act of appropriation. The technosciences aim at appropriating the world in order to disclose its instrumentality, developing knowledge of the instrumentality of the world in terms of the alethic modalities of appropriation. Contemporary critics of technoscience argue that it is predicated upon an objectification of Nature as something of inferior status and instrumental value to human beings.[41] However, this point of view about technoscience also treats it as if it were merely a positivistic instrument for the domination of Nature, but it also needs to be recognised that the human mastery over Nature is the goal of science, which is premised on the societal gamble in the ability of science to pacify Nature as the end result of a historical struggle. Even though these critics raise important points, they represent Nature as something passive and amenable to human domination. However, once we remind ourselves of the power and capriciousness of Nature, then we can recognise that technoscience is engaged in a confrontation with Nature as something aloof, looming, and terrifying that not only resists and thwarts human efforts to control, manipulate, and predict it, but is also capable of destroying whole cities and many thousands of lives as the result of an earthquake, millions of lives during an epidemic, or perhaps causing mass extinctions due to global warming or a meteor collision with the Earth. The goal of technoscience is the domination of Nature, but that goal is premised upon a primordial and anthropological fear of the capriciousness and power of Nature. It is a based on a confrontation with an active Nature, which is not readily pacified, and, therefore, it is the challenge and task of technoscience to replace the natural world with a safer and controllable artificial world. The objective of technoscience is the replacement of the natural world with the technological society, within which human beings can adopt the role of mastery and stewardship over the domesticated natural environment, but this state of affairs is more science fiction than fact. Whether or not human beings can pacify, appropriate, and transform the natural world into something of instrumental value is something that is very much open to question. It is for this reason that technoscience is an experiment, which is far from being completed, into the human productive capabilities to disclose the instrumentality of natural beings, as a means by which human beings can confront and pacify Nature. Once we recognise the experimental nature of this confrontation with Nature, then we can see that the societal aspiration for the innovation of the biotechnological capacity to remove randomness and inherited ills from embryonic development, supposedly creating "better" human beings, is itself a manifestation of the societal gamble and is itself an experiment in how to understand the mechanisms of life by constructing a "better" society. However, without any critical evaluation of the eugenic vision of society and humanity entailed by such a project, it remains nothing more than the irrational fear of the capriciousness of Nature and the faith in the goodness of the technological society. Before embarking on the construction of a brave new world for our great grandchildren, we should take pause and question whether such a world would be truly good for them. This debate should run deeper than evaluating the risks inherent to biotechnology and whether we should "play God", because we need to confront the vision of society (or the absence of such a vision) that the advocates of biotechnology propose. Of course, whether or not the biotechnologists can fulfil their promises is very much open to question.[42] But, we need to go further than this. We need to question whether they should fulfil their promises before they try to. Should they actually be permitted to try to construct the society that they envision? What vision of human well-being and utopian society do these advocates promise? Would such a society actually promote human flourishing or diminish it? What does human flourishing entail? And, if these advocates of biotechnology do not provide any satisfactory answers to these questions, then we need to question the rationality of their aspirations.[43]

Karl Mannheim warned that we are facing a profound danger that our society was heading towards a future in which it will be devoid of all ideological and utopian thought.[44] In this future, everything would be a repetition of the past and there would not be any possibility of transforming social reality and human nature. This would bring about a static state of affairs in which human beings would become objects. The irony would be that after a long historical struggle, when human beings reached the highest state of awareness, productive power, and creativity, at the point when human beings will be almost free of historical and material conditions, with the loss of ideological and utopian imagination, human beings will lose the will to transform social reality and thus lose the ability to understand history. However, even though I am critical of the positivistic interpretation of modern science and technology, I am also critical of the humanist criticism of modern science made on the presumption that positivistic technoscience is meaningless for the human lifeworld. This criticism misses the mark in one crucial respect. Humanists are quite right to criticise the instrumentalist interpretation of technology that positivism presupposes, on the basis that any uncritical and unthinking relation with technology endangers human freedom, but the humanist position has largely neglected to examine the extent that the positivist position is based upon a moral stance in favour of human freedom from the constraints, limits, and capriciousness of Nature. The positivist sees human freedom as being contingent on the acquisition of knowledge of natural laws and the enhancement of technological power achieved by rationally implementing that knowledge. The criticism that the positivistic sciences have discarded philosophical inquiry into the nature of meaning and truth is well founded (replacing concepts with formula, questioning with technique, and reflection with logical analysis), but the fundamental transformation of existential meaning that has occurred through the positivistic sciences has been largely ignored. It was through the scientific world-picture, within which magic, demons, mystery, occult powers, and oppressive dark forces had no reality, that reality was represented as knowable through cognition and material practices, as something that human beings could set forth upon as an ongoing project of improving human existence. Far from being meaningless, each and every confirmation of the predictability and utility of the reality disclosed through positivistic science is a reaffirmation of the meaning of the scientific world-picture and is a validation of the faith in the societal gamble. Each and every experiment, observation, and calculation is a contribution to the refinement of the faith that it is possible to construct a technological society within which human beings would be free of fear and ignorance. Every subjective experience of the natural world was represented as anthropomorphism, as the psychological basis of all myth and delusion, whereas positivism was represented as the development of bounded technical rationality as being the progressive movement away from the madness of the medieval world of supernatural forces, ghosts, spirits, demons, devils, magic, and providence. The capriciousness of Nature, lacking any spite or maliciousness, is simply the unfortunate and seemingly random consequence of natural laws and mechanisms upon human life due to the human inability to foresee, avoid, or prevent such events. Consequently, the societal gamble is a response to an anthropological fear about human impotence and vulnerability in the face of the capriciousness of Nature. Scientists are engaged in the technological imperative as a moral imperative to liberate human beings from the evils of the natural world and improve the human condition. This has profound meaning for the phenomenological character of the human lived-world. The drive for certainty and control operates upon technological innovation in order to transform "the unknown" from being the origin of mystery and humility into the possibility of future opportunity and emancipation through increased power. Modern science was founded upon an ideological struggle for the creation of a means to achieve utopian visions of the human lifeworld as an ultimate end of a historical and materialist confrontation with Nature. Every technical and scientific activity is an ideological and concrete contribution to the possibility of a future realisation of a utopian vision of a liberated human existence, in the sense that the ideational and semantic content of scientific discourse can be said to entail a confrontation with Nature in order to realise a vision of future society that is better than the "natural state". Both modern industrial capitalism and Marxism presupposed this fundamental representation of modern science and technology as being the objective interface for the human confrontation with Nature. Materialism positioned and represented objectivity in relation to the practical successes of science, as the basis for understanding the form and content of rationality in terms of technical rationality. Furthermore, even though capitalism and Marxism opposed one another on the question of what the specific institutional form and content of the technology society should be, both presuppose the rationality of the societal gamble in the construction of the technological society as a means of emancipating human beings from the limits and capriciousness of Nature. Henceforth, the political development of society was designated as a technological and economic problem, rather than a philosophical problem, which would be solved by constructing the technological society. However, the societal gamble entails a vision of what the technological society and the human good life should be, and, therefore, if we wish to rationally understand the origins and directions of the construction of modern society, then we must critically examine that vision. We should philosophically examine the rationality of the vision upon which the development of modern society is founded. If we find that there is an absence of any such vision, then we will be compelled to critically examine whether the development of society is rational and conducive to human well-being.

As Mumford argued, the Scientific and Industrial Revolutions were ideological revolutions within which machines were the link between the Universe and humanity, the basis for rational calculation, and the producer of human goods. For the scientists and industrialists, this ideology was akin to a religious faith. However, as Mumford put it,

> The decay of this absolute faith has resulted from a variety of causes. One of them is the fact that the instruments of destruction ingeniously contrived in the machine shop and the chemist's laboratory, have become in the hands of raw and dehumanised personalities a standing threat to the existence of organized society itself. Mechanical instruments of armament and offence, springing out of fear, have widened the grounds for fear among all the peoples of the world; and our insecurity against bestial, power-lusting men is too great a price to pay for relief from the insecurities of the natural environment. What is the use of conquering nature if we fall a prey to nature in the form of unbridled men? What is the use of equipping mankind with mighty powers to move and build and communicate, if the final result of this secure food supply and this excellent organisation is to enthrone the morbid impulses of a thwarted humanity?[45]

Numerous technological catastrophes and failures have demonstrated that scientists and engineers do not quite have the control and understanding of technology that they often represent themselves as possessing.[46] In fact, given that technologies can only be understood in hindsight by observing how they transform and are transformed when implemented into the world, the whole process of technological innovation and development is an experimental process. The irony of the technological society is that, in the rush to achieve mastery over Nature, it has created interdependent complexes of technological systems that are so complicated, vast, and heterogeneous that are beyond comprehension and control. As this whole society is founded on a perpetual and anarchic process of technological innovation, it is inherently experimental, unstable, and unpredictable. In the passionate drive to escape the capriciousness of Nature, human beings have become subjected to the capriciousness of an artificial world of our own making. Furthermore, particular developments in *praxis* do not necessarily reflect any inherent potential for human emancipation, but they conceal the authentic possibilities of human emancipation because they reproduce and limit the human evaluation of the possibility of alternative truths. The invariance of technoscientific knowledge and activity are not to be represented as proximities to objective truth, but, instead, are to be represented as the consequences of the invariance of the social trajectories and goals of the whole technoscientific enterprise. Without any rational account of the rationality of these trajectories and goals, the realist and positivistic philosophies of science amount to little more than the irrational justification of the internal logic of a faith in the construction of the technological society as being good for humanity.

In my view, Marcuse was quite correct to argue that, according to this analysis, the modern relation between human beings and Nature is one of "controlled desublimation" and "the pacification of existence". No longer is Nature mysterious, feared, worshipped, or offered supplication in dread of its capriciousness or wrath. In the modern world, the capriciousness of Nature is something to be dealt with, predicted and avoided, prepared for, as a technical problem. Even "the wrath of Nature" manifest in the harmful consequences of pollution, deforestation, or global warming is treated as if it were simply a causal consequence of our technological interventions, while the solution to any problems caused by technology is presupposed to be a technological solution. However, adopting a Freudian analysis, Marcuse argued that "the Destruction Instinct" has driven the technological conquest of Nature, at least in large measure, and this instinct is, in the last analysis, "the Death Instinct".[47] In as much as it seems driven to destroy the basis for life on Earth, I do not deny that the urge to conquer Nature could be based on "the Death Instinct", but it also seems to me to be equally important to emphasise, once we recognise the metaphysical foundation of the technological society, that this urge to conquer Nature is also based on a sublimation of technology and a idealisation of human beings as "modern man". "The Creation Instinct" also drives this urge, in at least equally large measure, and this instinct is, in the last analysis, "the Life Instinct". This is not a "conquest of transcendence", as Marcuse put it, but is an attempt to construct the transcendence of "modern man" over all existence, akin to the mythical efforts of the architects of the Tower of Babel, in the face of an all powerful and wrathful God. It is a profound metaphysical act of negation and creation, unconsciously reproduced in every innovative technical design and operation, which celebrates the birth, survival, and progressive future of "modern man". To participate in the technological society is a profound moral and ideological act. The ordinary, everyday thoughts and actions of every individual human being in the technological society are the particular instantiations and affirmations of the universal "modern man" that, due to ingenuity and foresight, became the Promethean conqueror of Nature and Lord of the Earth. The pacification of existence does not involve a conquest of Nature, as Marcuse argued, but involves a substitution of a technological society for the natural world. This involves a historical transformation of society in order to make Nature of little or no regard or importance (except as an object of beauty or wonder, as landscape or spectacle). The human struggle for liberation from the constraints and evils of the natural world has taken the form of the creation and naturalisation of an artificial society as a replacement that suits human beings better. This transcendence over Nature is a glorification of modern man, not as a conqueror of Nature, but as a being distinct from Nature, liberated by his own ingenuity from natural necessity. The confrontation with Nature is not one of human beings trying to defeat Nature; it is in fact the other way around. The confrontation with Nature occurs when natural events and processes *remind* human beings of the existence of Nature, as something that is not controlled by human will. It occurs when Nature capriciously demonstrates its power over human artifice. The pacification of existence is actually the historical process of building the technological society as a shield against the capriciousness of Nature. Its success would not involve defeating Nature, but rather creating an island within Nature that allows human beings to ignore Nature. The technological society is the means by which Nature is to be relegated to history and placed outside of the human realm. Hence the escapism characteristic of modern life, that Marcuse criticised, is in fact inherent to the construction of the technological society. Modern life is inherently one of escaping from the possibility of any possibility of a confrontation with Nature. It is about the creation of a safe world in which human beings can forget about disease, death, decay, and spontaneous destruction, into which, from time to time, the capriciousness of Nature erupts and thwarts this forgetfulness, in the form of hurricanes, heat waves, blizzards, earthquakes, outbreaks of new diseases, and even rust on metal, or cracks in the pavement. These events show the technological society to be fragile and incomplete. Once we recognise this, it becomes pointless to criticise the technological society on the basis of its instrumentality or one-dimensionality. Due to its moral and ideological basis, through propaganda operations of its widespread media, all such criticisms of the technological society will either be absorbed, as having been effectively countered by the self-evident goodness of the technological society, or the will be dismissed as being nothing more than the ravings of fantasists that have no bearing on the real world. Hence, criticisms of the technological society must engage with its metaphysical foundation. In this way, we can question the rationality of the societal gamble by demanding an examination of the vision of "the better future" that current norms and practices supposedly are directed to build the conditions for its possibility. It is through such an examination that the absence of any such vision can be shown to entail an irrational faith in the societal gamble. Thus, every act of labour and sacrifice, in the faith that these acts contribute to the creation of the possibility of building a better world for humanity, noble that such acts may be, are nothing more than the irrational conformity to a visionless and vacuous faith in the goodness of an unimagined and unexamined form of the future construction of society. It is this irrational conformity to a visionless faith that is the psychological source of the individual's susceptibility to totalitarian propaganda and the willing submission to the technological and capitalist economic imperatives of modern society. In order to effectively counter this susceptibility, we need to raise consciousness of the absence of vision and call for the critical articulation and development of the vision for the better world towards which our efforts are supposedly directed. In the first instance, we need to question the rationality of the faith in the societal gamble. Secondly, when there are clearly defined goals presented as entailments for a better world, we need to critically examine whether and why these goals are really desirable as societal goods, what the possible implications of their achievement could be, and whether we wish to commit ourselves to these implications. Thirdly, we need to develop a fundamental and ongoing critique of the social norms of "goodness" that are involved in the societal gamble. Without this deep critique, we could not hope to develop any rational understanding of our lives in the modern world, and how the purposeful transformation of the lifeworld forms the rational basis for labour.

5
Labour and the Lifeworld

In many respects, we are what we can do. Technological innovation transforms human agency, it shapes our present and future possibilities, and embodies and transforms the ideals and values of society. The operation of the technological imperative upon the organisation of all plans and activity is directed to take us nearer to the realisation of the success of the societal gamble, the removal of the evils of the natural world, and the achievement of a paradise on Earth. The drive to innovate, bringing novel inventions and new transformative powers, is inherently a moral drive within the technological society to improve and perfect it. When modern philosophy attempts to articulate and judge these ideals and values in a critique of technology, it should attempt to articulate and judge the ideals and values of society, critically examining our assumptions and presuppositions, to prepare a way for the rational evaluation of the vision for society, in relation to alternative visions of our future possibilities, capacity for rational understanding, happiness, and well-being. It should attempt to present a different sort of critical rationality from the dominant instrumentalist conception of technical rationality, not only to disrupt or escape from the autonomy of the technological, but also to clarify whether technical rationality is rational in a broader, philosophical sense. The philosophical task of putting forth alternative forms of critical rationality is possible through a critical reflection upon alternative visions of the ideal lifeworld and its relation to science, technology, ethics, power, and the fragility of human life. It is through such a reflection about the character of human existence, our aspirations, potential, and possibilities for change that we can rationally question, understand, and articulate our ideals, values, and conceptions of human happiness and well-being. What kind of world are we trying to make? Why will it be good for us to live in such a world? If we wish to understand the ways that technology mediates and shapes intentionality in the modern society, transforming that society and how we see its possibilities of development, then we need to move beyond the positivistic assumption that technology is simply a means to an end, and, instead, examine how technology is experimentally embedded in the ongoing development of the ontology of social being, how it relates to labour, and how it substantively transforms the human lifeworld. Of course, even if we accept that technical rationality, agency, and technoscientific knowledge are historically contingent, it does not follow from this acceptance that their successful dissemination and reproduction are simply based upon the successful dissemination and reproduction of consensus, a view often attributed to social construction theorists and postmodernists. This view tends to overly focus on the contingent and artificial construction of scientific facts, as if they were purely textural or discursive processes of exercising or establishing authority. However, while there are considerable critical insights into the nature of institutionalised social power and authority offered by social theorists, we also need to examine the processes of the artificial construction of the technological society as an ontological and practical experiment that is directed in accordance with a vision of human well-being and the good life, in response to an imperative to create a better world devoid of human suffering, miserly, and struggle. Once we recognise that technical rationality is itself contingent upon the societal gamble in its goodness for human well-being, then we are no longer compelled to accept the Enlightenment faith in the universality of this form of rationality. It becomes open to question. Moreover, once we also recognise the ontological and practical importance of this faith for our society, then we can critically address the way that it is central to underwriting the sharp distinction between practical and theoretical reason that allows technoscientific knowledge and discovery to be represented as being rationally applied to the problems caused by the deprivation of human passions, appetites, and needs.[1]

The recognition of the contingency of rationality does not necessarily lead to the rejection of the practicality of technoscience, providing that its substantive and experimental character is also recognised. The rationality of an evaluation of the applicability of any theory or practice should be itself evaluated in terms of the coherence and cogency of the rational evaluation within socially situated activities and projects in accordance with explicit goals. But, it is quite impossible to critically evaluate those goals, without an explicit account of human well-being and the good life, and, hence, provide a rational account of those goals. Scientific realists need to explicate an objective, universal account of human well-being and the good life, rather than merely assert an equation between truth and the good. Otherwise, it is quite impossible to evaluate the rationality of this assertion because such an evaluation would itself presuppose metaphysical assumptions that would require justification. Scientific realism is premised on the metaphysical assumption that it is possible to discover a neutral means to achieve cognitive transparency and conceptual correspondence with objective reality, and also that it is good for human beings to discover these means, but whether or not that is actually the case cannot be determined as an *a priori*. The presumption that the experimental sciences provide the neutral and beneficial means to achieve the self-evident goods of scientific knowledge and technological power are based upon the metaphysical interpretation of a grand social experiment. Thus, scientific realism is contingent, experimental, and provisional, but cannot be rationally founded until the technological society has been completed and our descendants are confronted with its final reality. Scientific theory and practice can be understood in relation to how it is integrated within the phenomenological reality of the lifeworld, not just in the pragmatic sense of how it provides the basis for technical action, but in terms of its political and economic role in organising human labour, intentionality, and participation in the construction of the technological society. Every scientific discovery and invention to this progressive development, no matter how small or specialised, can be represented as a contribution to the overall, incremental liberation and empowerment of humanity, but the uses and meaning of scientific discoveries and inventions do not remain fixed and defined at the walls of the laboratory and the pages of scientific literature. As they are disseminated throughout the wider world, these modes of use are innovatively transformed and extended as they are implemented into the wider world. The meaning of the products of scientific activity are transformed as they are embedded and integrated within the cognitive, discursive, and technological practices of the everyday lifeworld. They are also transformed as they are utilised for political, commercial, and military purposes within the institutions and policies of states, corporations, and armed forces. These dimensions of appropriation and usage are complex, pluralistic, intimately related, and frequently incoherent. It is evidently the case that the experimental sciences are instruments for the discovery and production of novel powers for whatever purposes and challenges the wider world demands, as well as being means to satisfy human curiosity about the natural world. New technologies can lead to unforeseen consequences and we can, in our ignorance, cause radical changes that we cannot predict or control. Just as the ancient agricultural practices radically changed the world, as human beings deliberately felled forests to clear land to plant new crops or farm domesticated animals, bringing new possibilities and opportunities into the world, they also brought unexpected disasters and problems. As well as all the complicated social changes brought by agricultural practices, including the possibility of empire building and the development of the feudal system, deforestation led to increased incidents of flooding and desertification, the spreading plants and animals led to the spread of new diseases, and monoculture increased the yield of the harvest at the expense of increasing the chance of famine when the harvest failed due to extreme weather or blight. Each and every technological innovation increases the level of ambiguity in the consequences of its implementation in a highly complicated world. It increases this level of ambiguity because the total meaning of any technology is not simply defined by the intentions of its inventors, but is defined in terms of the totality of uses to which it is put and its political, economic, and environmental consequences (which generate their own ambiguous and unforeseeable changes in the value and usage of any technological innovation). Once we recognise that the construction of the technological society is an ontological experiment in understanding the world by changing it, in order to achieve some ideal society in the future, we are compelled to accept that we cannot know what we are doing in advance. Today we are still struggling to understand and predict the climatic and social consequences of the Industrial Revolution. We cannot hope to be able to predict the implications and consequences of contemporary technologies, such as computers and biotechnology. All technological innovations are experimental interventions into the development of society. However, to understand the meaning and value of any technological innovation, we need to recognise that the goals of such activity have always been the acquisition and achievement of power – and as history shows us, the question of the meaning and value of technology depends on whom it increased power for. If we aim to develop modern society into a just and egalitarian society, then we must address the distribution of power within society. But, as I shall argue below, this is not simply an ethical imperative. By distributing power throughout society, allowing widespread participation in the decision-making process about whether and how to develop and implement any novel technology, the level of societal diversity and plurality of participation increases, which increases the capacity of society to survive and respond to unforeseen consequence, perhaps even anticipate them, and also broadens the bounds of rational decision-making by increasing society's capacity for lateral thinking and creativity. As I shall argue, this is only of any real benefit to society if there is widespread communication throughout society that allows people to learn and respond to the ideas, practices, and experiences of other people, through open and pluralistic interaction. In other words, the health and creativity of society is optimised if it is a pluralistic, participatory democracy. By optimising the level of plurality and diversity within an open, democratic society, it increases the chance that some members of that society will discover happiness and the good life by experimenting and learning from the experiments of others about how to live life well. While we may well remain sceptical about the possibility of achieving universal happiness, the health and creativity of society is optimised when we agree on universal rights for people and communities to explore a plurality of possibilities on how to live life well. Simply put, pluralism and diversity within society increases the chances of someone somewhere doing something right, it allows society to develop a stock of distinct contingencies and strategies which can come into their own should unexpected changes occur in the world, and allows society to benefit from the widespread distribution of diverse skills, ideas, and experiences of its citizens. Pluralism and diversity increases the societal capacity to respond and adapt to unexpected ontological changes, either through natural events or technological innovation, and therefore increases the health and creativity of society, providing that power is widely distributed throughout society. A genuine democracy can only emerge if we all face and struggle with its ontological possibilities, as if the meaning of our lives and the future of humanity depended on it.

Labour and the ontology of social being

Lukács argued that labour is a category of the ontology of social being that is complex and can only be understood through its relation with the overall character of social being, as well as in relation to the other categories, such as communication, cooperation, organisation, and production.[2] Labour cannot be understood in isolation or as an abstraction, as simply an economic commodity. For Lukács, labour emerged from the organic struggle for existence and all the steps of its development are products of social activity, which are understood through the ontological processes of the production of use-values. In this way, labour produces the conditions for its own reproduction and development (including specialisation and division of labour), which are causally and objectively understood in terms of the structure of society, rather than the inorganic properties of materials. From the onset, Lukács was critical of the cosmological interpretation of teleology that pervades the history of philosophy, and, instead, analysed the teleology of labour in terms of the conscious positing of a goal for both the overall process and its detailed steps. In relation to this goal, every participant in the labour process is objectively understood and organised in terms of its use-value.[3] Human beings project purposes onto the organisation of material practices, and, hence, labour is premised upon teleological positings of the modes of organisation of social and material practices, in order to realise social purposes materially. It is the ongoing teleological positing of labour that endows social being with autonomy from the organic conditions of human existence, while the teleology of labour is realised in the material transformation of the world in accordance with the posited purposes upon which labour is planned. This material trans-formation produces something new in relation to the natural world, such as a house, for example, which, due to its purpose, is objectively distinct from being merely a sum of the materials from which it is comprised. Labour brings new social possibilities into the world, by real-ising teleological positings in material combinations. Human beings are able to take these new social possibilities, and further produce more new possibilities, endowing new material combinations with new properties, in accordance with the new functions that these combinations have in relation to new teleological positings. Of course, as Lukács was aware, the ongoing development of the experimental sciences provides new possib-ilities and, as such, they are a condition for the realisation of certain teleological positings, say building a flying machine (which remained a dream from the Ancient Greeks to Leonardo Da Vinci and afterwards, until the Wright brothers actually built and flew in one), but due to the active ordering of the whole labour process in terms of the significance of each element of that labour process, as a use-value, natural beings are not understood as they are in themselves, but understood in terms of their significance within the labour process. Thus Nature is only under-stood in relation to labour as a category of social being, and, hence, human beings only relate to Nature through labour as being a means to satisfy ends, which, of course, reveals the conception of the world and the properties of manufactured objects, as well as the determination of the best means for the satisfaction of teleological positings, to be subor-dinate to social being. The properties of natural beings, such as stones, are determined and understood in relation to their appropriateness for labour.[4] Natural beings are transformed by the preparatory processes in relation to the teleological positings that direct the whole process of planning practical activity, and the objective properties of natural beings are understood in terms of a sequence of consequences that are identified and structured in relation to social practices. Thus any object is understood in terms of its function within teleologically posited labour, and also in dialectical relation to an imagined set of alternatives, as an object of conscious reflection on the nature of reality. As a consequence of this, the various natural sciences, such as the experimental sciences, are dependent upon the development of practical experiences, methods, results, and representations of labour, and their ontological connection with everyday life and social needs. The representation of Nature is transformed through specific conscious acts of the planning of the material practices arranged to discover natural mechanisms and laws, in accordance with how those practices are already understood in terms of a reflection upon the means and ends of practical activity. This results in a dialectical conception of objectivity, which is determined by specific modes of human activity, and, thus, is changed in relation to changes in how they are represented, organised, and connected with specific goals. Labour is dynamic and complex, dialectically emergent from produced modes of social being, in conscious relation to imagined possibilities and alternatives, in accordance with planned, practical activity.

For Lukács, labour is a natural activity for human beings, hence the boundary between the social and the natural may retreat with the increased complexity of practical activity, whether the process is that of making a stone axe or building a nuclear reactor, but Nature never fully disappears. However, causal accounts are produced through a practical engagement with materials, as a labour process, and, thus, the infinite possibilities of the properties of any natural being are reduced and selected in relation to the teleological positings implicit to that labour process of investigation. Every experiment is conducted on the basis of achieving a generalised theoretical evaluation of a labour process that is set in motion by purposefully grouping materials and practices in order to isolate and establish a set of interactions in terms of a hypothetically posited causal relationship, which is tested by evaluating its value for future practice. While it is possible to represent realised teleological positings as being those permitted by natural law, this determination of a causal account is itself derivative from the labour processes, itself based on the teleological posting of discovering those laws through experimentation upon material combinations. If these laws are expressed in mathematical formulation, then they must be an ontological interpretation as a physical process, in terms that are translatable and connected to practical activity (such as technical operations or possibilities of measurement). These laws are themselves abstractions of the representations developed in order to explain and quantify the consequences of the experiment in terms of possibilities, actualities, and limitations. The meaning of these laws is defined in terms of their use-value within the labour process they abstract and for the further realisation of new teleological positings that they suggest.[5] The natural object in itself and the natural object as an object for labour (with corresponding use-value, for something else) are heterogeneous. Thus the experience of the naturalness of an object often emerges when the object is in opposition to the teleological positings that direct the labour process within which the object is given meaning. An example of this would be the corrosion of iron. Efforts to understand this opposition are labour processes posited as the means to discover the means to remove the opposition, and, consequently, the opposition is itself understood in terms of imagined alternatives and possibilities, such as iron that does not corrode. Another example would be the way that iron resists manipulation, which is understood through the labour processes of overcoming that resistance. At no stage is the natural being encountered, as it is in itself, but it is always understood in relation to practical activities, such as measurement, experimentation, control, checking, repair, and so on. In this way, as Lukács observed, the conscious reflection upon the nature of reality is based upon an engagement with materials in relation to decisions between alternative possibilities. The object as an object of consciousness is a product of labour, available for labour, while the subject becomes a subject in relation to the attitude of transforming the objective world. The transformation between possibility and reality is mediated by social activity, such as planning, discussion, and modelling, and is very much dependent on heterogeneous criteria, such as economic or technical criteria. The final form of the transformation cannot be considered as being based deterministically on the possibilities and the means to realise them. The socially mediated goals of labour have a crucial role in the transformation between possibility and reality, given that they shape and direct the organisation of the whole labour process, as well as having to accommodate to decisions made in the particular concrete contexts of the conditions and consequences of actually doing the work. Thus the actual realisation of the goals of labour is neither absolutely free, nor is it causally determined, but is mediated, produced, and developed as a complex social being that transcends both the subject and object. It remains something over and above the components from which it is comprised, as something new that cannot be reduced to material composites, due to the inherent operation of human decisions and imagination on the goals to which practical activity is directed, nor is it an isolated product of absolute mind, given its alethic connection to physical reality through material practices. Freedom ontologically emerges in society in relation to alternative goals to which human activity potentially could be directed to realise, induced by social needs, in order to change the conditions and concrete actuality of social being, in relation to an imagined horizon of the potentiality of social being. This means that freedom arises due to the mediation and interpolation of teleological positings between social needs and their satisfaction. Alternative developments of the realisation of social needs are always possible, due to the alternatives that confront us at each and every stage of decision regarding the possibilities of labour available to us and the goals that we should strive for in order to organise the process of realising the satisfaction of those needs, as well as the processes by which those social needs are understood as needs. Every decision between alternatives forms the impetus of the ongoing development and differentiation of social being, while the realisation of these decisions, through labour, permits the further possibility of new alternatives and decisions. Human beings are confronted with decisions about what the character of social being should be, and, thereby, human beings are free to change human nature, given that social being is over and above the constraints and determinations of its organic and inorganic conditions upon which its possibility depends. The way that human beings adapt to a changing world is premised on a whole complex of decisions regarding what are possible and potentially realisable teleological positings for labour, what are the best potentials for social being, what human beings ultimately desire to be, and what kind of world human beings dream of living in. Freedom emerges from the choice between alternative dreams and visions, as well as the way that these aspirations are realised through labour. This process cannot be understood in terms of biological instincts or natural laws, but rather in terms of struggle for self-mastery within which social being becomes dialectically transformed as the constantly changing site of that struggle, against which the natural conditions of social being disappear further and further into the background, as labour transforms natural beings into objects understood in terms of their use-value and exchange-value, neither of which are intrinsic properties of the natural beings. Scientific knowledge or modelling of natural causality may well provide predictions of the consequences of our choices, but they do not provide the basis for making those choices. Moreover, once we realise the close relation between science and labour, then we can appreciate how the scientific knowledge and models developed and differentiated through labour are the products of a process of refining and innovating labour. This means that freedom is enhanced by scientific knowledge and modelling, once we recognise that scientific truth is bound together with technological power, and we equate freedom with the possibility of choosing between alternative goals that we have the means of realising through labour. This conception of freedom places necessity and freedom in dialectical relation to one another, rather than an antithetical relation, as assumed by the traditional idealistic and deterministic philosophies.[6] Human freedom emerges in the social struggle to use the possibilities of labour to overcome our organic constraints and limitations, in order to dialectically transform and master the material conditions of our social being in accordance with our teleological posited vision of our potential. Freedom consists in the mastery of labour: the self-mastery of human being.

Even the most rudimentary material practices require cooperation between people, and therefore the teleological positings of labour are always the positing of goals for other people. Cooperative social practice involves a division of labour, requiring a structured sequence and combination of planned material practices, and, therefore, labour is directed towards transforming social being, as well as transforming natural beings into objects of use-value and exchange-value. This twofold process of transformation involves the production of a total connection of subject–object relations that has primacy over the relations themselves. The meaning of any particular subject–object relation can only be understood in terms of the overall teleological positing of the total connection of subject–object relations, and, as such, any element of the labour process can only be understood in terms of its purpose within the total definition of that labour process. All the elements of the labour process are synthetically brought together and given meaning in relation to the complex of social being, and can only be understood in the context of an ontological analysis of social being, in terms of their purposes within this complex. As social being increases in complexity, then the development of new labour processes and complexes of labour processes develop into a system of complexes and processes, within which each complex, process, and relation becomes increasingly defined in terms of its relations to the system. And thus the system acquires an immediacy and supremacy over its components. In this way, the system forms even more autonomy from its organic and inorganic conditions, in such a way that specific relations, processes, and practices become generalised and fixed as an ontological basis for the system. For example, geometry was developed and differentiated from practical activities, such as measuring and building, and achieved an autonomy from those practices as a system of abstract relations of quantities and proportions. This allowed geometry to operate in science as the basis for generalised abstractions, experiences, and observations about Nature, without reference to the practical activities from which it was developed and differentiated.[7] In this way, the practical activities of chemistry can emerge from the magical practices of alchemy, while developing autonomy from those practices, independently from the truth or falsehood of their interpretation.[8] This is also evident in the development of modern mechanics from Aristotelian mechanics. Human beings develop and differentiate ways of understanding the relation between the human subject and the world in such a way as to develop a sense of autonomy and independence from the conditions of its development and differentiation. This sense of autonomy allows conscious reflection upon the systems emergent from social being to be represented as an unmediated, primary experience that corresponds to some objective state of affairs or internal categories of universal logic. Lukács considered this be the problem of false consciousness in that it ignores the complex developments, differentiations, and relations between theory and practice that were instrumental in the production and reproduction of this representation of the human subject and the world as object.[9]

Following Engels, Lukács argued that language, for example, arose as a means to satisfy organic needs, relating to danger, food, sexual desire, and so on, but rapidly developed and differentiated in levels of abstraction to such an extent that the abstractions become posited as independently existing objects.[10] Language is no longer bound up with the concrete particularities of the biological life process, conditioned by immediate organic needs and instincts, but achieves a level of generalisation that allows expressions and descriptions to be transferable between contexts, different groups of phenomena, and different subjects. As Lukács argued, once expression and description become transferable, then they become distanced from the objects they express and describe. This allows the subject to be distanced from the object, while language is represented as the medium between subject and object. Hence, once the objects of description and expression become increasingly abstract, then language can be represented as the means by which human beings are able to obtain mastery over the object by making it comprehensible and communicable. Language allows the development and differentiation of new relationships of mediated ends and means (each satisfaction being attained through labour, being something that is already represented in terms of its own objective nature) to be represented as the common possession of society that provides mediated means to satisfy immediate ends. Lukács used the example of cooking and eating meat to make this point: while boiling or roasting meat is clearly a mediated, social process, it is the case that the act of eating cooked meat is represented as being as immediate as eating raw meat from the carcase of a freshly killed animal.[11] According to Lukács, while the biological basis of human life remains unchanged, how that basis is mediated, satisfied, and transformed depends upon social processes, to the extent that this transformed mediation becomes the basis upon which human beings identify their needs and how to satisfy them. This social basis for life develops and differentiates autonomously from the organic conditions of life, even if human life cannot be ontologically separated from the organic conditions for sustaining and reproducing it, and how those organic conditions are represented, comprehended, and communicated also develops and differentiates autonomously from those conditions. In other words, while human beings have immediate inorganic and organic needs, we do not develop a conscious relation with these needs, but develop and differentiate a socially mediated conscious interpretation of our needs, described and expressed in terms of abstractions and generalisations, which are purposefully organised in relation to the complex of social being. However, when this socially mediated consciousness is taken as an immediate consciousness of our natural being, then the teleology of labour and language is represented as if it were the deterministic teleology and language of the natural order, which becomes represented in terms of the interaction between natural causality and intuited truth that becomes revealed to pure contemplation or immediate perception of the system of the world.[12] As a consequence, the historical materialistic rejection of teleology simultaneously became an affirmation of determinism, which resulted in a reified abstraction of the conditions and consequences of labour in terms of natural mechanisms and laws. The dependence of labour upon the complexity of the ontology of social being became concealed by this false consciousness (made possible by suppressing the historical conditions and choices made to create the particularities of this mediated consciousness); hence, particular historical, concrete actualities of social being became idealised and reified as the being of "Universal Man" in an immutable cosmos, within which labour became implicitly subordinate to an absolute mind or deterministic laws.

According to Lukács, the solution to this problem of false consciousness requires reference back to theory and practice in their concrete
appearance in the processes of labour, in relation to social being, in order
to show the real and authentic role of teleological positing in materially changing reality, through its dynamic postulation of a possibility
towards which labour can be organised and directed towards realising
in cognitive and material combinations. This shows how teleology and
causality can be related through mediated, complex processes of labour
and material practices, but without referring to either cosmological teleology or deterministic causality. In this way, the actualities and potentials of human being become unambiguously understood in terms of
social being, and social theory and practice can only be fully related in
terms of their reciprocal developments and differentiations as elements
of one and the same complex of social being. The antithesis between
true and false is such that it is only revealed through labour as being the
realisation that any organisational incoherence between the teleological
positings of any labour process will in all likelihood lead to its failure.
However, given that the posited goals of labour inexorably involve the
participation and interests of other people, based on decisions between
alternative possibilities, it is impossible to understand labour in terms of
causal chains, and, therefore, it is an error to consider the heterogeneous
positings as being scientifically understandable in terms of causality in
the same way as natural science understands natural processes. This is
even more salient once we recognise that natural, experimental sciences
depend on specific theoretical interpretations of labour processes in
order to understand natural processes in terms of causal chains. Scientific
theory is raised to a higher level of generalisation as the science in question develops autonomy from practical activity, but this is not possible
without mediating the interpretation of Nature in relation to expectations, intentions, and interventions, which themselves are specific
modes of social being. The goal for any labour process exists (in the
imagination) prior to its realisation; given that the process by which this
realisation is brought about is directed, step by step, in accordance with
the posited goal, then the meaning of the posited causality is such that
the causal sequences are understood, selected, and exercised, in relation
to an anticipated future. Even when the participation of Nature is understood as that which works or resists by itself within the labour process
(the material response to human interventions), this is not a spontaneous cause, but is a response to purposefully directed interventions,
made in relation to the ordering of the techniques as a series of means
that are posited to achieve specified goals. At no stage is it possible for

us to encounter Nature, as it exists independently of human intentionality, expectations, representations, and practical activity. Even though
Lukács adopted the traditional Marxist interpretation of labour as being
an immediate form of the human confrontation with Nature, once we
look past this dogmatic adherence to the orthodox posture, in terms
of Lukács' own theory of the social ontology of labour, including the
mediated nature of the natural sciences, then we see that his description
of the social–natural relation is closer to my description of society as an
attempt to create an artificial world as an improvement over the natural
world, as a means of liberation from the limits of organic being and the
capriciousness of Nature. This attempt finds its clearest expression in
the drive towards the autonomy of the complex ontology of social being
from its organic and inorganic conditions, which increasingly disappear into the background, as social being creates the means to distance
itself from Nature. A posited future that is governed by a technological imperative, which teleologically directs this drive to appropriate
and transform social being in order to improve society in accordance
with an imagined good life, is a completely artificial construction of
an idealised aspiration to realise the best of all possible worlds that is
evaluated in terms of its opposition with representations of the state
of natural being, in terms of animality or organic conditions. It is on
the basis of this implicit, ontological "ought" – this ongoing striving
for this posited future ideal – that the contemporary state of affairs is
judged in terms of an imagined potentiality, from which an idealised
projection of the objective standards of rationality is posited as a moral
imperative upon all social being. The ongoing material development and
differentiation of society is an expression of this moral imperative, and
it is along these lines of development and differentiation that objects
achieve their identity in terms of their use-value. In all but the simplest
labour processes, the social complexity of labour leads to the projection
of objectification over other people when they are organised and objectified in terms of their use-value, for the purpose of the development
and differentiation of society through practical activity. However, due
to the production of new use-values through labour, the objectivity of
use-value is made in relation to the labour process, which, of course,
means that it has a historical, dialectical relation with social need. As
new use-values are produced, new possibilities arise, and there are new
social needs, which cannot be reduced to organic and inorganic relations, or understood in these terms, even if they are an ontological
condition of social needs. The emergent autonomy of social being is
the historical realisation of the specifically human, understood in terms

of the capacity to reproduce developed and differentiated practices and relations.

The economic and political developments and differentiations of social being are qualitative achievements of new systems of ontological complexity, even more removed from the organic and inorganic conditions of labour, upon which power relations and exchange-values can be developed and differentiated from use-values, through the mediation of the system, which, in turn, transforms them and the possible outcomes of conscious reflection upon them in terms of their value for future use. As these forms of greater social complexity arises, a whole new form of objectivity becomes possible, which makes political and economic activities possible as mode of social being placed in connection with its newly emergent and objective properties. Once social being reaches a sufficient level of complexity, then economic and political systems become normalised, represented as the rational consequences of the objective properties of social being, which become "second nature" for human beings. Economic and political activities become represented as the fundamental expressions of human essence, though which we identify and understand ourselves in terms of our function within the system, outside the personal relations between family and friends, outside the privacy of self-reflections. The problem arises when, once economic and political systems become objectified, normalised, and "second nature", an antagonism arises between the objective, law-like representation of economic and political rationality, and the phenomenal forms of human life that make it possible. This antagonism between the deterministic characterisation of the economic and political basis of society and the purposeful nature of social being manifests itself in an ontological contradiction between the representation of the development of society as being governed by laws and the purposeful, pluralistic, decision-making, participatory nature of human beings. This contradiction leads to its resolution through the representation of ethical or moral reasoning, aesthetic judgements, as well as all sensuous activity, as being subjective and ephemeral in the face of the necessities of objective, economic, and political reality. However, this resolution is one of mere appearances (a mask, so to speak) that conceals the class-based hierarchy of values, which are profoundly imposed upon every member of society, in order to represent the values of the social elite as being the objective necessities of the whole of society. Given that the ontological complex of social being is dynamically given substance through the material realisation of values through teleologically posited labour, the conflict between classes is inbuilt into society and its subsequent developments and differentiations. This substantively directs and limits the possibilities for consciously realising alternative means, values, and goals in the structural transformation of the further developments and differentiations of social being. This directs and limits the possible decisions about which values and goals to materially realise in future teleological positings of labour. These directions and limitations become reproduced and sustained in traditional representations of exemplary modes of social being (such as in mythical heroes and social ideals), while, at the same time, being reproduced and sustained as duties, laws, and interpretations of social needs and the best means to satisfy them. The substantive interrelation of these representations in social practice allows specific ideals, values, and goals as being represented as historically objective and progressive. This implicitly involves a suppression of the plurality of ideals, values, and goals, as well as the heterogeneous structures and differences between people, in order to synthesise an homogeneous totality of social being, as reproduced and represented in social practices, institutions, and relations.

Technology and the lifeworld

Albert Borgmann used the example of the TV-dinner to show how technology has transformed contemporary life.[13] He used the notion of a "device paradigm" to articulate the fundamental drive towards efficiency within the technological society, but at the cost of reducing the richness of social life. The microwaved TV-dinner emerges ready for the individual, quickly and effortlessly, in contrast with the laborious and time-consuming preparation of a traditional family meal. Of course, a scientifically manufactured TV-dinner can efficiently provide all of the individual's nutritional requirements, without involving all the social complications of bringing family members together, at the same place and time, and preparing and cooking the ingredients to produce the same meal for individuals with different tastes. Borgmann observed the way that the device paradigm transforms eating into an efficient, technical operation of consuming and controlling nutrients and calories that places the biological need to eat over the social meaning of the family meal, which thereafter becomes a luxury or an antiquated and tolerated inconvenience. The ritual of the family meal becomes an obsolete means of satisfying the needs of everyday life. Borgmann lamented the loss of the ritual reaffirmation of family unity, this "focal thing", as he termed it, because, without the effort to maintain such social relations, they cannot survive the functionalising effect of modern world and, as a consequence, much of the richness and meaning of social life will be lost. Borgmann assumed that technology reduces individual involvement with other human beings to a bare minimum, and transforms technique and control into the highest values. In his later work, Borgmann concentrated on the computer.[14] He was concerned with the way that the person as a focal thing has become a commodity delivered and mediated by computers. He argued that this new way of relating has weakened connection and involvement while extending its range. As he put it,

> Plugged into the network of communications and computers, they seem to enjoy omniscience and omnipotence; severed from their network, they turn out to be insubstantial and disoriented. They no longer command the world as persons in their own right. Their conversation is without depth and wit; their attention is roving and vacuous; their sense of place is uncertain and fickle.[15]

Don Ihde developed Borgmann's point further when he described the disembodiment of agency through the use of computers and the Internet. Given that the body is characterised through agency within the technological society, then, by using the Internet, the body is actually dissociated from its extended agency.[16] This agency remains embodied with the body–computer interface, demonstrating the radical transformation of the body that has taken place. It becomes transformed into an agency defined in terms of its remote functionality and speed, both becoming tacit forms of knowledge through the acquisition of skills.[17] The body becomes synonymous with the mode of interface and its enhanced powers, but this transformation is itself situated within a lifeworld in terms of tasks, procedures, and techniques, as well as liberating and empowering aspects of intentionality. But this empowering relation between the body and the computer dissociates the agency from the body as subject, transforming it into a set of operations and appropriate responses, reducing the social richness and sensorial character of the lifeworld.

However, contrary to Borgmann, technology functions as a substitute for human involvement in circumstances where that involvement was already weakened or lacking. It substitutes technique and control as the highest values as a consequence of an absence or impoverishment of other values, and, as such, does not erode our social values at all, but simply acts as an alternative. The example of the TV-dinner shows how technology fills the lacuna that occurs when traditional relations become obsolete or neglected. Indeed the demands of modern society on human relations are immense, but the disintegration of the traditional family is not caused by modern technology. It is more the case that modern technology transforms the lifeworld when it is adopted as the means by which previously impoverished situations can be remedied through technology. The pernicious aspect of modern technology is not that it causes the disintegration of social relations, but rather that it conceals that fact that these relations have disintegrated. We do not need to adopt a reactionary critical attitude to technology *per se*, but, instead, need to closely focus on our neglected social relations and examine how we have allowed technology to act as a substitute. On this account, it is not the case that technology reduces the meaning and richness of the lifeworld, but it is more the case that technology has filled the lacuna within an already impoverished lifeworld and, thus, obscures the extent that richness and meaning has been lost. It conceals the absence of alternatives by allowing us to function effectively as individuals within a society in which technology becomes the medium for social interaction. Its success as a substitute conceals the inherent loss of meaning and value in our social relations because it allows us to ignore the fact that our social relations and values have become eroded, but it was not technology that eroded them in the first instance. The use of the computer and the Internet is another case in point. It seems that for many of us the computer offers a means to communicate with people at a distance, sending them emails, chat messages, photographs, files, and so on, whilst we also maintain face-to-face relationships and involvements with others. For most of us, our computer-mediated conversations are neither more nor less lacking in depth and wit than our letters or spoken conversation. Our attention is not made more roving or vacuous by our use of a computer to search the Internet than it is when we browse library shelves or flick through an encyclopaedia looking for something interesting to read. For most people, the computer is another means for communicating over distance (faster and cheaper than the postal service, saves on paper too), gathering and organising information and entertainment. The incorporation of the computer and the Internet in our everyday lifeworld does not erode the pleasure in having dinner and conversation in a fancy restaurant with one's lover or spouse. It does not reduce the pleasure of talking until dawn with an old or new friend. Playing a computer game with one's child does not erode the pleasure of playing ball games in a park. Email and chat services do not replace the time spent with friends and relatives. However, in fact, it seems to me that Borgmann has taken as the norm a minority of people for whom social intercourse is a profound problem that is solved by using computers to communicate. Hence, he missed the way that the computer operates as a substitute whenever interaction was already lacking. Computers and the Internet have helped people for whom social intercourse is difficult or lacking, but, because it has helped these people, the pernicious aspect of the computer is that it allows such individuals to function in the modern world without examining or dealing with their social difficulties and isolation. It prevents them from confronting their difficulties and fears by acting as a substitute, therefore, removing the need for such a confrontation by providing an ersatz remedy. For such individuals the computer acts as a medium in which the individual can forget their alienation and control their social interactions from a distance. In this way, the computer reinforces their difficulties and isolation, acting as a substitute for face-to-face social interaction, and concealing the alienation of individuals by providing a solution to their social isolation and phobias. It is for such a person that the virtual network represents a reduced and abstracted social life, which leaves them insubstantial and disorientated when severed from the network.

While there is considerable merit to Ihde's analysis of the reduction of human experience of relating to others via computers, contrary to Ihde, human contact is not only measured by full body co-presence, something that we more often than not take for granted, but, rather, is brought to the fore during the absence of full body co-presence. Both the telephone and the Internet (chat rooms or email) are cases in point because they provide technological means by which the absence of full body co-presence between family and friends can be artificially bridged. Contact becomes precious and based on personality rather than full co-presence because, while this non-sensory mode of communication remains something of a compensation for the absence of full body co-presence, it also enhances one's feelings for the absent friend, relative, and/or loved one, in a way that is over and above one's emotionality when the person is present. Absence may well make the heart fonder, but this fondness is emergent when one simultaneously has a trace of the presence of the other, especially when it highlights their absence. In this sense, the dissociated agency of the body becomes a means by which the technology becomes the medium of sensuality and aesthetics, which transforms them when it is the means to make them possible.

Through such activity, the human subject is immersed in the technological society as an agent and a resource for other agents. Human beings are both the object and subject of technology. We use tools and instruments; we are also tools or instruments for others. The intentionality of the subject is one that is mediated by technology, as it extends towards a horizon of ends, while being situated and directed in accordance with the intentionality of others, in which case, the subject is objectified as an agent or functionary within a system. Technology not only extends the body (providing it with enhanced physicality and power), extending the subject's horizon of possibilities and intentions, but it also constrains and directs the embodied subject in accordance with the technical framework of associated means and ends. Intentions are transformed into abstract objects that can be passed from body to body when those bodies acquire new techniques. Technology mediates and defines how we perceive ourselves, how we perceive others, and how we perceive others as perceiving ourselves. It situates us within a system as a person defined by agency and functionality, capable of achieving objective goals, using objective techniques. The body not only is empowered and extended through technology, but it is also identified as a being in terms of its technological empowerment and extension, as an object at the disposal of the system. It is in this sense that technology can be considered as having a transformative effect on the lifeworld. It transforms how we experience and interact within the world, in our everyday activities, as well bringing new content into the world, mediating and transforming the lifeworld within which we relate to each other and the world. The modern aesthetic experience of technology is not simply the experience of the beauty of machines, represented in terms of their power and progress, but, much deeper than this, for already initiated and trained technical practitioners, it brings forth a sense of its naturalness, simplicity, and autonomy. The tacit understanding of the capacities of the machine, realised in every expected operational shift in performance, converges the motility of the human body and the productivity of the machine into a harmonious flow of effects from technical activity. It is through this trained perception, through the mediation of representations of the machine performativity, that the technical practitioner is able to make observations of the causal effects of an underlying objective reality at work in the responses of the machine to operational interventions. Thus the machine can be represented as the means by which new experiences are made possible, allowing the human body to be identified as the conscious conduit for natural forces to be released through the design, construction, and operation of the machine. As Mumford put it,

Science gave the artist and technician new objectives: it demanded that he respond to the nature of the machine's functions and refrain from expressing his personality by irrelevant and surreptitious means upon the objective material. The woodiness of wood, the glassiness of glass, the metallic quality of steel, the movement of motion – these attributes had been analysed out of chemical and physical means, and to respect them was to understand and work with the new environment.[18]

Technologies do not just provide new experiences; they transform how we understand what it is to experience and how to communicate our experiences, while they also act upon the lifeworld, transforming it in unexpected ways. Once established and institutionalised, as a system within the ontology of social being, technologies can achieve autonomy of operation that is often distinct from the intentions of its designers and engineers. It is commonly known that when technologies are implemented in the open and messy wider world, then they produce unexpected side effects and uses beyond the intentions of the designers. These include environmental effects (such as pollution, ecological destabilisation, or aesthetic transformations of the landscape), social effects (such as unemployment, new possibilities for crime, or unforeseen economic impacts), and existential effects (such as new possibilities for agency, new representations of what it is to be human, and transformations in the way that human beings relate to each other and the world). The definition of any new technology can only be understood once it is implemented and integrated into the world as a total set of functions, effects, and transformations, and, therefore, it remains ambiguous and experimental as it becomes integrated into the current political, economic, and technological systems, transforming and being transformed during the ongoing process of combining the new technology with the pre-existing infrastructural complex and lifeworld within which it is situated. In this sense, once we define any technology in terms of what it does (rather than only in terms of what its designers and engineers intended), then we see that, in fact, the term "side effect" is a bit of a misnomer. There are only effects. Thus radioactive and thermal pollution are not "side effects" of nuclear power, but are effects and are part of its total definition as a technological process of providing electricity. Its capacity to cause social anxiety, awe, fear, and nationalistic pride are effects, rather than "side effects", that are also part of its total definition. The meaning of any technology can only be understood in relation to its total definition, which of course is never realised because we still are not aware of all its effects and future technologies will transform them. For example, we are yet to be fully aware of the effects of the practice of burning coal to generate heat. We have yet to provide a total definition for this ancient practice, let alone all the complexities of the development of nuclear power. In itself, this should provide something of a cautionary signal about the innovation and implementation of radical new technologies, such as biotechnology, when we consider our ignorance about fossil fuel burning and its effects on the atmosphere, human society, industry, politics, and economic relations. Consequently, when we allow modern technology to become a substitute for absences and lacunae in our social interactions within our lifeworld, then allow our interactions and relations to become mediated by a system for which there is no possibility for predicting or controlling the consequences of that act of surrender, we should take pause for critical reflection. Given that we simply cannot know what will happen as a result of developing and implementing any technological innovation within the world, the substitution of the system for interpersonal interactions involves the ontological displacement of our rational deliberation upon the meaning and purpose of our social being into that of objective and impersonal technically rational choices regarding upgrades, desired functions, and how to achieve increased technological power and efficiency. This involves the total surrender of our free relation with technology because it substitutes the objective relation of the system for our authentic and personal relations of the lifeworld. As Mumford put it,

> The displacement of the living and the organic took place rapidly with the early development of the machine. For the machine was a counterfeit of nature, nature analysed, regulated, narrowed, controlled by the mind of men. The ultimate goal of its development was however not the mere conquest of nature but her resynthesis: dismembered by thought, nature was put together again in new combinations: material synthesis in chemistry, mechanical synthesis in engineering. The unwillingness to accept the natural environment as a fixed and final condition of man's existence had always contributed both to his art and technics: but from the seventeenth century, the attitude became compulsive, and it was to technics that he turned for fulfilment.[19]

The meaning of any technology is emergent and transformed through this process of integrating it into the existent systems, and the consequences of this process of integration cannot be determined in advance, simply because its ontological possibilities and functions are emergent properties of the process of integration. The technological mediation and systematisation of social being transforms the lifeworld into an objectified system that is inherently unstable, uncontrollable, and autonomous. Once human beings have surrendered themselves to the system, then it acts as a mediating substitute for the lifeworld and the natural world by appropriating in accordance with both the objective structures and the emergent properties of the system. It was for this reason that Habermas rejected the possibility of the realisation of the ideal harmony between human beings and Nature.[20] He realised that modern science and technology were premised upon instrumentality, and, therefore, it is impossible to base modern science and technology upon a communion with Nature. The logical structures of science and technology are grounded in "purpose-rational action", in the form of work aimed at controlling the world or communication in pursuit of common understanding. He represented communicative action as being able to limit technical action to facilitate the development of the complex interactions required by modern society, in a rational and progressive process of enhancing human freedom and producing a better society, while technology was represented as a means of achieving control and work by objectifying Nature in terms of its instrumentality.[21] In his critique of modern society, Habermas made an analytical distinction between systems (media-regulated rational institutions) and the lifeworld (the sphere of human interactions) in order to critically examine the way that the criteria of systems are imposed upon the lifeworld in order to make the lifeworld technically rational and efficient in the terms of the system. Habermas developed a concept of the lifeworld as being the sphere of everyday communication and activity, in which we as subjects actively pursue shared understanding and purposes, within which everyday life, education, and public deliberation occur. He was concerned about the extension of technical rationality and its criteria for efficiency into the sphere of communication and interpersonal relations. His concern was whether it is possible for us to overcome the constraints of modern society and develop free and rational communications. This provided a distinction between function and meaning, in order to show how the technological society does not sustain such a distinction, but operates on an ambiguity between function and meaning that is implicit to our everyday systematised lifeworld. Function provides meaning, and *vice versa*, in a technological society within which functionality has aesthetic and moral value. In order to be effective, technology requires its implementation and integration into concrete contexts of the systematisation of social relations and practices. Thus, technology becomes the systematic objectification of meaning and values, and society becomes the institutionalisation of techniques and procedures to achieve those values. By empowering, transforming, and innovating modes of agency, while integrating them into a system, technology has a substantive effect on meaning of the lifeworld, but that meaning is ambiguous and open to radical transformation in accordance with the systematic integration of that technology into the lifeworld.[22] The "technization of the lifeworld" replaces the social interactions of the lifeworld with a system of media (such as the exchange of money, the exercise of administrative power, or the utterance of stereotypes to transmit performances of the expression of belief or desire), which transform and objectify social interactions into techniques and products.

For Habermas, modern society imposes an imbalance between the system and the lifeworld because it objectifies and mechanises deliberative and purposive-rational action (transforming it into a technique to achieve pre-specified ends) and reduces the need for linguistic agreement on what constitutes value and meaning in the face of everyday ambiguity. The medium for the ontological organisation of social and individual action becomes a system of power and monetary exchange for which institutionalised business, administration, and law become the rational forms of calculation, control, and decision-making. Modern society diminishes the role of communication within human interaction because the market mechanism imposes a satisfactory result and suppresses the need for discussion about what constitutes a mutually satisfactory result and why. Media become a substitute for communication, which co-ordinate individual action in large-scale projects that are instrumentally directed towards pursuing individual success in the acquisition of money and power. Media-steered information replaces genuine communication with objectified and reproducible "shared beliefs", rather than reach them in the course of linguistic exchanges. Thus it reduces the depth and richness of meaning that can be achieved through human interaction by transforming it in terms of its function, technique, and objective results. Habermas' whole project was rooted in a critique of the positivistic understanding of reason and its historical realisation in a technological society. Our capacity for reflection should not be supplanted by the positivistic extension of technically exploitable knowledge.[23] Hence, he focussed on models for "purposive-rational action" and his criticism of modernity's counterpoints in the Enlightenment effort "to develop objective science, universal morality and law, and autonomous art according to their inner logic".[24] The specific forms of science and technology depend on institutional arrangements that are historically contingent and variable, but their basic logical structures are grounded in the very nature of purposive-rational action. The positivistic effort to demarcate politics, economics, science, and technology from one another – in each case by methodologically separating an approach to the world from all others with which it might otherwise be linked, thereby turning each approach into an autonomous pursuit that acts on a reduced aspect of the world – effectively reacts back on the world itself to reduce the world to a series of unlinked disclosures revealed by distinct techniques. Through the totalitarian mediation of the techno-logical society, which encompasses all relationships and intentionality, human beings are transformed into universal instances of rational and technological agency, identical to one another as instances of a species obtaining its evolutionary momentum through the medium of techno-logical innovation, liberated from their existential isolation as cooper-ating participants in a technological framework. Our consciousness of the technological society is mediated and replaced with the scientific world-picture within which the very structures and mechanisms of that society reflect the economic and political relations of the system, as if these were the consequences of natural law. It is this mediation that allows our conformity to the system to be self-consciously represented as the only logical consequence of rational deliberation.

Of course, the societal gamble entails the belief in the human ability to rationally use artifice for the purpose of enhancing the possibility of self-preservation, but it is important to note that there is quite a specific sense of "rational agent" that is entailed by this belief. It refers to the material, embodied "rational agent" that is capable of using its own skills, knowledge, and intelligence to save itself through the manipula-tion of causal powers in the natural world. The individual is represented as having been thrown like Adam (or Robinson Crusoe) into the natural world, and it is only his or her technical abilities and intelligence that will save his or her humanity. The dignity of the "rational agent" is one that is defined in terms of technological agency; "reason" is iden-tified as the ability to think in accordance with the logic of bounded technical rationality. The inability to do this, whether by choice or infirmity, would be one that would threaten the "rational agent", risking dissolving it into the inhumanity and indignity of acting in accordance with the organic state of animality that modern civilisation has striven to replace with the technological society. This would not only leave the body increasingly vulnerable to the capriciousness of Nature, but would allow a conscious oneness with Nature and the corresponding state of animality, as a channel for that capriciousness to flow through, by awakening primordial states of irrationality, passion, and instinct. The fundamental dread of modernity is that, without the framework of bounded technical rationality, the body would become capricious to the extent that it would become one with the primordial being of the natural world. It is quite ironic that the positivistic rationalisation of the domination of civilisation over the natural world, with its absolute denial of any reality beyond the manipulation of materials in accordance with the rational discovery and utilisation of natural laws, is founded upon a dread of the animistic and irrational nature of the natural body. The scientific objectification and rationalisation of Nature into a system, taken to the extreme of positivistic materialism and cybernetics, is as based upon a moralistic and repressive fear of the capricious and spontaneous nature of ontological being, as well as the wild and uncontrolled aspects of the world within which we exist. This fear is suppressed and concealed through the concentration of the positivistic sciences in the acquisition of information, the industrialisation and mechanisation of labour, the utilisation of the natural world as a resource, the constant affirmation of the representation of progress in terms of increased precision, power, and control, and the operation of the system. Thus the scientific experience of the world must be detached from the sensuousness and phenomenology of embodied experience in order to subjugate it to the objectivity of the technical and disciplined mediation of instrumentation, measurement, and calculation. Labour is reduced to being a systematised and rationalised mode of work, rather than a fundamental aspect of social being, within which the consciousness of the function of work is alienated from the social ontology of labour, once labour has become mediated within the system in terms of its exchange-value, quantified in terms of money and time, rather than its wider consequences and meanings for the ontological development and differentiation of the lifeworld. As Marx put it,

> Through the subordination of man to the machine the situation arises in which men are effaced by their labour; in which the pendulum of the clock has become as accurate a measure of the relative activity of two workers as it is of the speed of two locomotives. Therefore, we should not say that one man's hour is worth another man's hour, but rather than one man during an hour is worth just as much as another man during an hour. Time is everything, man is nothing; he is at the most the incarnation of time. Quality no longer matters. Quantity alone decides everything, hour for hour, day for day...[25]

The technological society and social power

As Langdon Winner pointed out, technological innovations are akin to legal and political frameworks in that they structure and institutionalise modes of public order that endure for many generations.[26] There are always alternative directions for technological research and development, and the decision regarding which direction is explored is a socially contingent and contextually situated process of establishing consensus and commitment. Once this decision has been made then it becomes increasingly difficult to explore alternatives. After any technology has been designed, developed, and implemented in practice, then it becomes gradually integrated into the technological infrastructure of functionally interdependent technologies, agencies, institutions, and organisations. It becomes an ordinary part of the everyday lifeworld and thus immediate and enduring resistance confronts the development of alternatives. The everyday lifeworld is constituted from and transformed by the technological choices and practices of others, both living and dead, and, in the most part, we have not personally participated in many of these choices and practices. Through the choices and practices of our ancestors, as well as distant contemporaries, often unknown to us personally, the social and private relations of the lifeworld are enframed (empowered and constrained) by a technological framework that preexists and transcends each and every one of us. Technologies structure and direct our modes of social interaction and participation, establishing and constraining opportunities for action and reflection, and direct how we live within the world and relate to our lives. Once we take this into account, then we can see that technosciences do not simply change society by providing new means and ends, but, rather, constitute and structure the ontology of society by fundamentally transforming what we are and how we relate to each other. The technological society is the systematic objectification of societal meanings and values embodied in machines, architecture, logistical infrastructure, political and bureaucratic institutions, and sciences, each associated with the reproduction of social power and authority. The capacity of the technological society to discipline and organise human beings is built into the mechanisms of policy formation and implementation, in everyday practices, as well as the scientific processes by which efficient functionality and productivity are represented and understood. This capacity is itself exercised and reproduced though the inscription of bounded technological rationality upon each and every specific policy and practice within the ongoing development and differentiation of systems that order practices and implement policies. Reform is itself transformed into a mechanistic process of adjusting the bounds of technical rationality and refining the representations of efficient functionality and productivity. Of course, in any industrial society founded upon the interests of an established oligarchy, the social elite finds their needs, desires, and ambitions satisfied by the structures and mechanisms of that society, which had been developed in order to produce, maintain, and reproduce that hierarchy for their benefit. In this sense, the individuals that form the social elite can be said to possess social power. However, these individuals do not have any choice in the means they must employ in order to reproduce the structures and mechanisms that maintain their status. Nor do they choose the needs, desires, and ambitions that society can satisfy. These are built into the structures and mechanisms of society. The individuals that comprise the social elite do not manipulate society in order to benefit from it; they must adapt to society in order to acquire their privilege. The capitalist elite may well posture as being the engineers of the future, but they too are conforming to the use of the instruments of power (media, machines, and weapons) in order to sustain and reproduce the inequalities upon which their class depends.

Langdon Winner's idea of "autonomous technology" – that technology is out of control, independent of human direction – has many parallels with Ellul's idea of "technique", as something totalitarian and irreversible, with its own autonomy and capacity for domination.[27] The technological society is becoming autonomous over and above the "will of its rulers", which increasingly has to conform to its imperatives – otherwise "the rulers" become replaced with someone else who is willing to conform – just as much as "the ruled" have to. The totalitarian mediation of the technological society encompasses the intentionality and relations of the capitalists just as much as the proletariat; therefore, even though capitalists clearly enjoy the fruits of the technological society (and the inequalities built into its structure and mechanisms), their policies, practices, and activities are controlled and directed by the very structures and mechanisms of the system upon which their privileged status depends. As Michel Foucault argued, power takes the form of self-control and adaptation; it is the product of the reproduction of particular discourse and practice in order to self-impose a regime of truth that informs what we should expect and choose, as well as how we should behave.[28] Power is constructed and functions on the basis of transforming other powers and their effects into a "tightly knit grid of material coercions" that becomes social embedded as "objects of power".[29] It emerges through social interaction and practice as a system within which the mechanisation of the means to reproduce power (and particular hierarchies) is based upon the institutionalisation and mechanisation of social interaction and practice. For example,

From around 1825 to 1830 one finds the local and perfectly explicit appearance of definite strategies for fixing the workers in the first heavy industries at their work-places. At Mulhouse and in northern France various tactics are elaborated: pressuring people to marry, providing housing, building *cités ouvrières*, practicing that sly system of credit slavery that Marx talks about, consisting in enforcing advance payment of rents while wages are paid only at the end of the month. Then there are the savings-bank systems, the truck system with grocers and wine merchants who act for the bosses, and so on. Around all this there is formed little by little a discourse, the discourse of philanthropy and the moralization of the working class. Then the experiments become generalized by way of the institutions and societies consciously advocating programs for the moralization of the working class. Then on top of that there is superimposed the problem of women's work, the schooling of children and the relations between the two issues . . . – so that you get a coherent, rational strategy, but one for which it is no longer possible to identify a person who conceived it.[30]

It is also essential to recognise that the inequalities of the structures and mechanisms of the technological society are not just inequalities between economic classes within the so-called "developed countries", or even between the developed and underdeveloped countries. Of course, the poorest peoples (especially the poor of the least developed countries) have the least influence on the development and implementation of technology, the conditions of their own labour, and the ends towards which technological innovation is directed, but even among the most affluent, women, children, and the disabled are also frequently marginalised by the technological infrastructure of our society. In order to successfully resist and address inequalities within society, it is futile to attack individuals and neglect to transform the mechanisms and structures through which those inequalities are reproduced. Thus, according to Foucault, power is not possessed by those who govern, and he was critical of the Marxist formulation of "the ruling class" and their analyses of domination, because "the main objective of these struggles is to attack not so much 'such and such' an institution of power, or group, or elite, or class, but rather a technique, a form of power".[31] By means of using an image of electrical current as an explanatory metaphor, Foucault described resistance as being the transformation and divergence of power through the expansive and pervasive circuits and relays through which power is transferred.[32] Existing forms of power are continually resisted by manifestations of other forms of power. It is only by becoming consciously aware of techniques, as forms of power embedded in structures and mechanisms, that people can change or withhold their participation in the reproduction of power, so as to achieve a reversal or redirection of power, rather than simply changing one social elite for another or generating endless cycles of resistance.[33] The democratic control of technology is possible provided that we stand up, act, and place technology under control by making decisions about the criteria by which we can rationally evaluate the development and implementation of any particular technology. If we wish to resist social inequality and transform the technological structures of society, in order to realise the potential for an egalitarian society, it is essential that we develop and continue sustained critiques of the structuration of social power within the technological infrastructure of society, in order to resist and redirect the forms of power reproduced in the mechanisms, structures, and institutions of society. This involves the development of pluralistic visions of society and human well-being, as well as egalitarian participation in the choice of which technologies to develop and implement in order to realise those visions. In this way, the rationality and aesthetics of technology should be bound up with social participation, community, and self-respect, as well as being central to the democratic structuration of the technological infrastructure and technoscientific endeavours within society.

Industrial capitalism emerged from the mass production capabilities of the Industrial Revolution. The laws of capitalist economics may well have been extended to cover every aspect and manifestation of social reality to such an extent that the whole of society is subjected and unified in accordance with the capitalist system, but the ability of capitalism to achieve this unification and universality was only possible because the development of the technological society empowered the premodern ambitions (such as the lust for power, wealth, and superiority) upon which modern capitalism depends as an economic imperative. Hence, the Marxist critique of modern society is inherently a critique of capitalism's corruption of the technological society by propagating profound social inequalities and dividing the proletariat into individuals (in order to suppress their class-consciousness and political power), but capitalism is only able to do this and be productively successful because the technological society is able to empower individualised workers in terms of a system of techniques and productive power. Whereas the ideology of liberal capitalism imposes the belief upon the worker that he or she only has economic value in terms of his or her labour – as a commodity – and that individual wealth was measured only in terms of the exchange-value of the products of labour, it was only able to successfully disseminate these representations, without destroying the cooperative productive base upon which capitalism emerged, because mechanical realist precepts underwrote a scientific world-picture within which technology is the rationalised transformation of natural and individual human capacities into techniques and, thus, is the fundamental representation within which the worker functions as a component – an efficient cause – without any awareness of the massive cooperative effort required for technology to function at all. Hence the scientific world-picture conceals the contingency and social ontology of modern technology. The transformation of labour into a commodity became the medium for rational human cooperation in a context in which the technological base of capitalism and the proletariat became one and the same. It is when the technological society – embodied in the proletariat – lacks self-consciousness of the imposition of the capitalist economic imperative upon its operation that productivity becomes synonymous with supply–demand and efficiency becomes synonymous with the reduction of costs. Within the capitalist system, the technological society is always simply a means to an end – increased wealth and power for the capitalists and wages and sustenance for the proletariat – but the proletariat has a unique position (being the embodied productive power of the technological society) in that it is possible that the proletariat can become conscious of the extent that it is the source of all productive power for both itself and the capitalist elite that exploits it. Hence, the Marxist argues that the liberation of the proletariat from this exploitation is inherently dependent upon its own self-conscious emergence as a class and that it is the productive power of society. What prevents the emergence of this consciousness? The mechanisation of labour has established a trajectory that began by reducing human skills in order to do away with them entirely. The motility and intentionality of the human body is being transformed from an organic mixture of improved and specialised activities into a set of machine performances with an increasingly automated social totality. However, instead of eliminating toil, transforming human reality into a utopian lifeworld, liberated from all fear and suffering, we are transformed into mechanised components within a total machine that transforms all our labour, leisure, and every daily activity into a predictable, controllable, standardised, and repeatable set of machine performances in accordance with their exchange-value. To the extent that human beings find escape from the capriciousness of Nature, we must submit to being controlled and ordered within the technological society. While society is under the dominance of the capitalist economic imperative, then each and every human being remains under the shadow of the capriciousness of capital. The premodern fear of the capriciousness of Nature has been transformed into the postmodern fear of the capriciousness of the technological society.

However, as Andrew Feenberg argued, the choice between alternative directions of technological development does not depend on technical or economic efficiency, nor on the intrinsic properties of technological systems, but on the socially established coherence between devices and the beliefs of the various social groups that influence the design process.[34] However, while I am in agreement with his sociological analysis about the contingency of technological design, implementation, and development, it ignores the extent that "the technical" is a subcategory of "the social" and how material practices and transformations can limit or empower social choices.[35] Both technical and economic rationality are bounded forms of rationality, in that what constitutes a rational choice emerges from a historical background of previous choices and trajectories, and, therefore, are socially contingent, but they are also limited by what is discovered to be the alethic modalities of material practices. While these alethic modalities emerge from the background of human activity, expectations, and assumptions about reality, they are not controlled or determined by human intentionality, in so far as their actuality is not circumscribed and limited by contingent human choices. It is this capacity for the consequences of technical activity to frustrate and surprise us that is a condition for the technical to transform the social, as well as being contingent upon it, in terms of non-linear, ontological relations. The identification, exploration, and representation of "the intrinsic properties" of technological systems are dependent upon the contingent construction of the system, while the success or failure of any experiment is based upon the coherence or incoherence of that system, in relation to its environment, which includes other systems. Consequently, any technological system cannot be understood only in terms of social construction of practices and discourse, or in terms of a correspondence theory of truth. It is an emergent property of its interactions within its environment, as well as the internal interactions of its components, that is contingent upon and conditions specific modes of social activity, for example particular acts of labour, intervention, and representation postulated to achieve specified goals, but it is not determined by them because these modes are themselves the consequences of an incomplete, experimental mode of social being that is itself uncontrolled and creative in its engagement and interventions within the world. Social being does not transcend the world, and any attempt to explain it in terms of any underlying objective social or material reality leaves us with an equally mysterious, ambiguous, and contingent concept of the social or the real. While I agree with Feenberg that technologies are underdetermined, the social processes of determining the limits and uses of any technology involve activities within the technical, as a subcategory of the social, in order to produce representations of the cognitive-material consequences, or machine performances, that emerge in response to experimental human interventions, developed to coherently understand these representations and activities in terms of a technological framework that precedes the experiment, which is used as a starting point for the societal implementation of that technological innovation in the wider world, within which the experiment is situated, funded, and directed as a means of producing prototypes. Feenberg's critique of the technological society was based on the way that technical rationality is based on a "decontextualisation of technology", which ignores the social meaning of technology in terms of its political value for the purpose of being able to publicly establish particular technologies as being the rational, efficient choice, and, therefore, the current form and content of the technology society is represented as being the only rational choice available to us. Feenberg was quite right to criticise this representation, but he assumed that its proponents are simply epistemologically mistaken about the neutrality of technical rationality. However, in my view, the situation is that they are politically in favour of the representation of technical rationality as neutral because they consider it to be essential to do so in order to preserve the *status quo*. Even though he was clearly aware that the social ontology of the technological society is structured in accordance with the interests of a social elite, Feenberg assumed that the social epistemology of knowledge is neutral, and, as a consequence, he argued that technological development is a contingent process, in order to show that the current form of modern society is contingent, without taking into account that the social epistemology that justifies determinism is implicitly bound up with the preservation of the current form of modern society. Neutral technical rationality is not a myth or an illusion; it is a tactic within a political process that aims to justify the continued use and further development of technologies that support the vested interests of a social elite that wishes to represent the technical infrastructure of society, upon which their power depends, as being neutral, lacking any rational alternative. It is pointless to merely deny the existence or truth status of neutral technical rationality; it is necessary to disclose its social function in sustaining antidemocratic power relations. Technological determinism does not ignore the plurality of equally rational alternative technologies – it suppresses them and justifies that act of suppression by representing it as a technical discovery of the most efficient means available.[36] Even though Feenberg was aware that technology is political, he ignored the extent that the philosophy of technology is political also. Technological determinism is an ideology bound up with political propaganda disseminated in order to represent irrational social choices as being the only possible rational decisions. Determinism is not "a species of Whig history", as Feenberg put it,[37] but is a rhetorical justification of a political process of closure designed to support the choice of one direction over another by claiming that it was the only logical option, while simultaneously controlling the whole process through which the information is presented and the decision is made. It is not sufficient to point out that we need to learn that the trajectories of technological development are contingent, if we do not control the process by which choices are made and justified, and, therefore, it essential that the public gains the social power required to be able to choose the framework within which decisions are made and how we can best choose that framework. What we need is a radical alternative to the current framework of decision-making, as well as the power to implement those decisions. Otherwise, apart from being aware of how cheated we are and how undemocratic our society is, we are none the better off.

Feenberg argued that technocratic ideology is effectively capable of silencing the call for widespread public participation in political decision-making decisions regarding the innovation, implementation, and development of new technologies. How does Feenberg explain the fundamental transformation of the lifeworld in order to explain the effectiveness of technocratic ideology? Feenberg's answer is surprisingly shallow. Following Bruno Latour, he points out that devices (such as an automatic door closer) enforce moral obligations in our society.[38] All mechanisms prescribe the values, duties, and ethics that humans have built into these devices and their contexts of use. In my opinion, this is correct, but it does not explain the effectiveness of the technocratic ideology. If technocracy is the use of technical prescriptions to conserve and legitimate an expanding system of hierarchical control, as Feenberg lamented, we need to explain the fundamental social transformation that empowered the technocratic hierarchy in the first instance. Feenberg did not do this – he has merely pointed out one of the mechanisms by which the empowerment is reproduced and disseminated. We need a deeper explanation. If technocracy is able to hide its evaluative bias behind the façade of neutral technical rationality, as Feenberg claimed, then we need to ask how this was socially possible. As I have argued above, once the rationality faith in the societal gamble became accepted as positivistically self-evident, then the legitimacy of any research project could be rhetorically supported upon representations of efficiency and social need. Once mechanical realism was implicitly accepted as the basis for knowledge of the natural world, technoscientists became the rational experts, those who had the knowledge and means to discover and implement the efficient means to remove lack. Those that paid the technoscientists and funded their work were able to decide what was lacking in society and, once they owned the means to satisfy this lack, it was in their vested interest to maintain this means and the lack. Further technological developments that removed the lack entirely were undesirable to this vested interest, at least, until it had made all the profits that it could, and, thus, alternative technologies were to be suppressed. It is thereby imperative that we bring the technological society under our reflective gaze and question its direction and content. This inherently involves questioning the rationality of the societal gamble, in order to rationally question the goodness of the technological society, and, if we do not like the answers we discover, then we need to transform our society into a more rational and better society. Of course, any rational evaluation of the societal gamble will presuppose substantive statements of what concretely harms or benefits human well-being, as well as an account of the good life. While there is an absence of any universally accepted account of the good life, there is a plurality of contenders, and it is the existence of this plurality that gives us a starting point to proceed rationally. The absence of any universally acceptable account, side by side with a plethora of contenders, is indicative of both human innocence and ingenuity. All such accounts form an experimental basis upon which human beings understand and narrate their lives. Thus every rational evaluation is itself based upon an experimental way of life. If we accept the objective existence of harms and benefits for human well-being, at least in the sense that some activities and conditions are good or bad for us independently of our knowledge and description of them, then the criterion for the rationality of any way of life becomes an *a posteriori* evaluation of whether or not human beings flourish or flounder by living thus. Such an evaluation must always remain pragmatic and tentative, given that future events may well lead to a reversal of fortunes and the need to re-evaluate our interpretations of any way of life. It is for this reason, if no other, that societal pluralism, in the sense of different communities and individuals trying different ways of living and evaluating their lives, can be considered to be beneficial to the health of that society because it limits the societal damage of any errors in judgement regarding the evaluation of the good life and how to live it, while it increases the chance that the good life could be achieved through trial and error. An inherent diversity in the decision about how to live life well and healthily is one that provides us, as a species, with a buffer against the capriciousness of Nature, in that any violent change in our environment is less likely to universally damage us as a species because its impact will not be universally felt. In fact, for some of us, it could well be an advantage. However, this benefit of pluralism can only be achieved if there is an effort to communicate evaluations of the good life across society in order for communities and individuals to learn from others, as well as help others learn how to live life well. Thus, pluralistic society is only beneficial for being pluralistic if it is democratic. It is to the question of how we can develop and implement a genuine democratic basis for the technological society that I shall turn to in the next chapter.

6
Into the Future

In Plato's dialogue, *Protagoras*, the chief interlocutor, from whom the name of the dialogue is taken, narrates the following tale. When the gods decided that the time had come to populate the earth with living beings, they entrusted Prometheus and Epimetheus with the task of producing them and providing them with suitable qualities. Epimetheus took over the concrete work, while Prometheus reserved for himself the right of supervision. After having wisely distributed among the different living species the characteristics that would enable them to survive and reproduce harmoniously, Epimetheus discovered – when the moment came to produce human beings – that he had already exhausted the natural qualities. So he was obliged to produce a being that was naked, weak, devoid of any special feature, and inferior to the animals. In order to remedy this oversight, Prometheus stole fire and the arts (that is, the principles of making) from Hephaestus, and he stole from Athena the arts of the intellect (that is, the principles of science and wisdom). These qualities were diversely distributed among human beings and they, by using them, were able to secure their superiority over the animals by producing artefacts and building cities. However, humans showed themselves incapable of living in communities, as they split into factions and fought with one another. At this juncture, Zeus, much concerned about the destiny of humans, punished Prometheus and charged Hermes to bring to humans the political virtues of justice and modesty. After these virtues were bestowed, humans were able to live a harmonious life in their cities. Obviously, this tale is a myth, but perhaps it is timely that we reflect upon how we need to exercise these virtues, harnessing our technological powers, in order to direct our efforts to building and living in harmonious and sustainable communities, exploring the diverse possibilities and alternatives that our imagination and ingenuity suggest, and

175

putting our technologies to work in helping us all live a good life. If the gift of Prometheus is forethought, then is it not time for us to use this gift to put forth our vision of society? If we do not want human beings to become powerless slaves to technological innovation then we have to subject technology to articulate critical thinking. Philosophy, in a general sense, needs to reaffirm itself as the primary mode of critical reflection upon the conditions, values, meaning, and ideals of our technological society. We need to philosophically scrutinise the vision of the world that we are working towards creating. Is it really desirable? Would it really be good for us to live in that world? However, we are neither titans nor gods. It is impossible to accurately predict the forms that technologies and societies will take in the future. We cannot know whether our current actions will be good or bad for us in the long run. We are in a state of innocence regarding the future, but we are not beasts that are only concerned about the present.[1] We think about the future and can try to find the best course of action to try to achieve our ideals and visions of how we should live. Every action is a gamble on its own goodness. It is for this reason that we should embrace social plurality and diversity because, once we accept our existential innocence about the future and the consequences of our actions, we also must accept that we simply do not know which course of action is for the best. We are all guessing. If all human beings are equally innocent regarding the future and the goodness of our actions, then we should adopt an egalitarian stance about human goods, values, and purposes.

Society has to accommodate different visions of the ideal society, which may stand in conflict with one another. The philosophical question 'what is the good life?' has a plurality of answers. It is thus better for society, as a whole, if its citizens try as many varied courses of action as possible and to seek to satisfy as many different goods, values, and purposes as possible. We need to explore alternatives if we are to maximise our chances of discovering good courses of action and to minimise our chances of making a terrible mistake that damages the whole of society. It is simply the case of not putting all our proverbial eggs in one basket. A pluralistic society has a greater chance of hitting on a good course of action, through its trial-and-error processes, involving as many people as possible in as many different ways of life as possible, than an authoritarian and dictatorial society which collectively follows the same path that was pre-emptively set down by a single individual or social elite. It is simply a question of the advantage of diversification over specialisation in an unpredictable and changing world. Once a pluralistic society, comprised of diverse communities and individuals exploring as many different ways of life as possible, has developed widespread possibilities for communication and debate among its citizens, then we are able to learn from each others' trials and errors, and debate and discuss how best to live our lives in the light of our collective experience. Such deliberation and reflection also needs to be pluralistic and diverse, incorporating alternative visions for society, as well as divergent, critical evaluations of the nature of rationality, all of which should be available to aid our decision-making processes about how try to live life well. Such a society is inherently a democratic society. Of course, in a democratic society, evolving in complicated and pluralistic ways, some individuals and communities are going to make some terrible decisions and, due to the capriciousness of Nature and the unpredictability of technological innovation, unfortunate events and consequences can interrupt the most carefully crafted plan. However, even if some individuals or communities fall afoul of their mistakes, natural disasters, or unforeseen consequences, a pluralistic society has a better chance of surviving and learning from mistakes and misfortunes because other individuals and communities will be trying different things for different reasons, and, therefore, the impact of those mistakes and misfortunes will be varied. By recording and studying its detailed history, as well as engaging in critical discourse with people from other communities, we will be able to learn from our whole society about our possibilities and limitations, about our experiments, about our failures and successes, and about how we can celebrate our differences as being our greatest source of creative power and our greatest asset for constructing a sustainable society. It will remain an ongoing, genuine, and mindful effort towards a sustainable and desirable life for as many people as is humanly possible, and that is perhaps as close to progress as it is possible to be, even if we all cannot agree on what a sustainable and desirable life would be. Technoscience needs to be integrated into society – not simply through the trial-and-error laissez-faire of use, consequence, and accommodation, but through an open, critical, and democratic evaluation of the vision for society as a whole. Such a public and pluralistic evaluation of the benefit and impact of technoscience should be made in relation to evaluations of human well-being and potential. When the direction of technological development and innovation is "governed" in accordance with the short-term interests of a social elite – without much regard for human well-being or potential – we find ourselves dominated by institutionalised technological practices that are socially damaging, unjust, inefficient, and environmentally disastrous. When the social evaluation of technoscience is restricted to whether it satisfies the short-term interests of an oligarchy, the development of the technological society will remain that of an inflexible, undemocratic, and totalitarian means of acquiring more wealth and power for a social elite. However, the democratic development of the technological society does not simply depend upon the egalitarian distribution of technoscientific knowledge and political power. It requires these, but it also needs flexible, democratic participation in the evaluation, implementation, and re-evaluation of criteria under which technoscience should be evaluated. Public participation should not be limited to merely making a choice between representatives because this will continue to limit decision-making to the narrow criteria of bounded technical rationality, political expediency, and cost benefit analyses. Public participation needs to be involved at all levels of decision-making, bringing a diverse and pluralistic stock of imagination, knowledge, experience, and values, in order to broaden the criteria of evaluation needed for a genuinely open, societal exploration of visions of the ideal society and the human good life.[2]

Decentralisation and democracy

Friedrich Hayek argued that a centralised system of planning would inevitably lead to totalitarianism and tyranny, and, therefore, socialism was a utopian delusion.[3] By collectivising power and putting it at the disposal of an authoritarian committee or leader, within a system that demands that all individuals concentrate all their efforts for the benefit of the whole society, the decision-making resources of the whole society would be concentrated in a small group of people. It is quite inconceivable that any small group of people would have sufficient intelligence and knowledge to amass a sufficient degree of foresight in order to adequately plan the construction and development of a society. It is also quite inconceivable that a small group of people would be able to anticipate and plan for every consequence of every possible event that could occur in the future. Centralised planning increases the societal vulnerability to collapse or disaster, and, therefore, it will inevitably result, sooner or later, in economic failure or the inability to cope with some unforeseen event. It would also inevitably lead to a complete loss of individual freedom. According to Hayek, only the worst elements of society, those that utilise deception to appeal to the lowest common denominator and basest instincts, are capable of gathering the massive support that is needed to govern the majority ("the docile and gullible", as Hayek put it) that have no strong convictions of their own, but will accept any system of values providing that it is "drummed into their ears sufficiently loudly and frequently" and it is based on ideas that take advantage of human weakness, such as hatred for an enemy or fear (of enemy infiltration or terrorism, for example). Thus, according to Hayek, the majority will be comprised of those that are most readily influenced by propaganda because they are possessed by vague and ill-formed ideas and they are most readily aroused by their emotions and passions. Hence, Hayek argued that when elected officials embark on a course of economic planning for the whole of society, often described in vague terms using populist rhetoric, then there will be an agreement on the need for central planning and an absence of any agreement regarding how to implement the plan. This, of course, is doomed to failure from the onset because its only criterion for success is that the majority has agreed upon the policy. Moreover, Hayek argued that, due to the diversity and plurality of the possible courses of action in society, it is impossible for democratic assemblies to function as planning agencies. Compromises only result in the construction of an unworkable plan or everyone being dissatisfied with the results. Ultimately, democratic assemblies need to delegate the task to experts or charismatic leader, and this will result in power residing in the hands of a single person or a few individuals. Hence, Hayek argued that the best way to limit and decentralise power is to distribute it throughout all the individuals of that society and allow competition between those individuals to provide solutions to our problems. Hayek argued that it is only by basing economics on decentralised competition can we hope to preserve individual freedom and optimise societal creativity, by utilising individual knowledge and creativity. He advocated liberal capitalism as the best form of social organisation because, providing that there is some basic insurance against the common hazards of life and a clearly thought-out legal framework, individual competition to acquire personal wealth provides the best basis for human efforts to be effectively co-ordinated without any intervening authority.

Putting aside that Hayek did not make it clear how this level of insurance was to be decided and distributed, nor who was to do all the thinking through of this supposedly clearly thought-out legal framework, my main contention with his thesis is with the fundamental idea that liberal capitalism provides the best ideological basis for people to organise their efforts. In my view, the problem with liberal capitalism is its unquestioned premise that a principle of laissez-faire competition will lead to decentralisation. Unfettered economic competition always favours those with the most wealth, resources, and power because it allows them the greatest access to technology, and to enjoy the benefits of an economy of scale and capital flight to regions with cheap labour. When empowered by the technological society, liberal capitalism will inevitably lead to an industrial capitalist economy based upon corporatism and globalisation, which results in increasing amounts of wealth and power in the hands of a small social elite that is capable of moving its capital to countries which provide it with the cheapest labour, while being able to provide the means to produce more goods, at lower costs, and transporting them across the world. Industrial capitalism is far removed from Benjamin Franklin and Thomas Jefferson's ideological visions of an idyllic nation comprised of liberal individuals engaged in competitive capitalism in order to increase their personal wealth and, thereby, the wealth of the nation.[4] Liberal capitalism may well have been a reasonable and achievable ideal when it was based on the craft-based workshops and small farm-based agricultural markets of postcolonial Massachusetts and Virginia, but when empowered by the technological society, liberal capitalism leads to the concentration of wealth and power in the hands of a few individuals because small businesses and farms are unable to produce goods cheaper than large businesses or farms, and, therefore, an economy based on competitive individualism will inevitably end up with a few individuals controlling all commerce, industry, agriculture, and mining. An economy based upon competition allows individuals with greater access to resources to be able to increasingly dominate the economy, which eventually results in an oligopoly planning the world economy and the directions of technological innovation. It may well be the case that more people in the so-called "developed countries" have access to cheaper goods, but the consequence of industrial capitalism is that the majority of individuals end up having the terms of their labour and material conditions of their life decided by the few. Once corporations acquire equal resources and power as nations, the only difference between the centralised state and a corporation is regarding whether the members of some executive committee for a state bureaucracy or the board members of a corporation make the decisions that affect millions of people who have no say whatsoever in the economic and political development of their lives. Either way, the concentration of power and the social stock of available imagination and creativity are reduced to that of a few individuals dictating the criteria for evaluating the development of the technological society. Once this is coupled with a political system within which those that can afford access to mass media and expensive political support are the only ones that have a chance of election to legislative bodies, then the process by which the legal framework and levels of social insurance are established are decided by the wealthiest members of society. When empowered by the technological society, liberal capitalism transformed into industrial capitalism and is as likely as any socialist state to result in a totalitarian system of centralised planning that is organised solely on the basis of preserving the *status quo*. The best conditions for propaganda are exactly those produced through industrial capitalism: where a massive media infrastructure is owned by a social elite with the vested interest in maintaining the *status quo* and is the main source of information and connection between individuals and government, which also has a vested interest in maintaining the *status quo*.[5] Under these conditions, the conformity of individuals to the *status quo* is the most likely result. Hence, contrary to Hayek, once we take the ontology of the technological society in account, then we can see that liberal capitalism is just as likely to result in totalitarianism and ineptitude as any other system and, consequently, is also the road to selfdom for the majority of people.

The contradiction between the capitalist economic imperative and the technological imperative is most evident in modern industrial societies that are considered advanced and prosperous, because, within these countries, a significant proportion of their populations lack access to education and health, modern housing and hygiene systems, civic security and legal support, nutritional and balanced foods, and opportunities for creativity and leisure, while a minority enjoy all the fruits of technological empowerment and are able to participate in its development. In order to maintain a sufficient proportion of the population for low-paid and low-skilled employment, in order to reduce labour costs, industrial capitalist societies must deny any real access to civilised and democratic life for a significant proportion of society in order to maintain a standing-reserve of workers. Thus, industrial capitalist societies must maintain a reduced knowledge and skill base of a substantial section of society, as well as limit the possibilities for wide democratic participation in the evaluation of the public goods and the development of the technological society, and, as a consequence, subordinate the technological imperative to the capitalist economic imperative. Despite all its promised competitive advantages of the self-regulating, laissez-faire market, industrial capitalism has led to the concentration of capital in oligopolies and monopolies, which, of course, creates an antidemocratic basis for the technological society and leads to all of the problems of centralisation that Hayek considered to be the inevitable consequence of socialism. Moreover, we need to rationalise consumption, rather than squandering precious resources to produce manufactured goods that we really do not need and often cannot afford, and, therefore, we need to address the extent that profiteering opportunists hoodwink us into participating in irrational and meaningless purchasing behaviour by equating the good life with increased purchasing power and credit. We need to recognise that the human good life depends upon much more than increasing our levels of expenditure and consumption. If we wish to rationalise the productive and consumptive base of our society, then we need to democratically decide the extent that localised industry can satisfy our needs, how we evaluate costs, and whom the "global free market" actually benefits. Do we really find it acceptable to constantly erode the rights and benefits of local workers simply because factories that utilise cheap labour in other countries can produce cheaper manufactured goods elsewhere? According to the dictates of the capitalist economic imperative, a global economy of acquisition of increasingly cheaper goods is a free and progressive society, but this evident racketeering is clearly for the benefit of the minority of the population that profits from pathological consumerism in one country by exploiting the unprotected labour force in another. Do we find it acceptable to exploit cheap labour in underdeveloped countries to provide the developed countries with cheap goods and commodities? We need to seriously question to what extent that an economic system that defines "competitiveness" and "efficiency" in terms of reduced wages and longer working hours is actually leading us in the direction of a free and progressive society, or whether it is a nothing more than a thinly veiled system of coercion and slavery. However, once we submit to the technological imperative, then we cannot accept the practical benefits of technology without also accepting its moral imperative of enhancing the material and aesthetic conditions for the whole of society. The social denial of this moral imperative leads society to fragment and degenerate into an irrational system of inequalities and exploitation, within which a small part of society benefits by manipulating and sabotaging the educational and psychological development of the rest of society, in accordance with the dictates of the capitalist economic imperative. Within such a society the societal benefits of industrial technologies are appropriated and, instead, it becomes a pernicious source of opportunism for the few. However, the unfettered development of the technological society, only under the technological imperative, is one that must eliminate social distinctions because its universal and collective goals are that of the empowerment of labour and the use of science to liberate human beings from the limitations of our material conditions. In order to maximise the efficiency and productivity of society, as a totality, this empowerment and liberation must be distributed throughout the whole of society. The technological efficiency, capacity for innovation, and sustainability of any society – as well as its ability to respond to unforeseen natural disasters and other unpredicted events – is dependent on its internal level of diversity, plurality, and cooperative responsiveness, which are all maximised when knowledge, skills, access to resources, and capacity for decision-making are distributed throughout the society.[6] The capitalist insertion of their pecuniary interests into the technological development of society necessarily limits (and even retards) the empowerment and liberation of society because it must disempower the majority of society disproportionately in order to protect the social privileges of the minority. Indeed, industrial capitalism provides large numbers of people with a wide variety of low priced, mass produced goods, providing that there are sufficiently large numbers of people coerced to work for low wages, but the true cost of that "benefit" is that an even larger and increasing number of people are unable to develop sustainable and democratic communities because the control of their economic basis is far removed from them. It becomes impossible for citizens in a community to plan the development of their lives if the technological and economic infrastructure of their community can be removed or transformed by people outside of their community, simply because it is more cost efficient to operate industry or agriculture in underdeveloped countries rather than developed countries. If the implementation and development of the technological society is dictated in accordance with the short-term economic and political interests of a minority, it is impossible for any community within that society to develop a democratic and rational basis for its own organisation and development, because it cannot plan for its future. This limits the bounds for strategic decision-making and, hence, leads to increasingly limited criteria for the development of society, which, of course, reduces the diversity and pluralism of society, the creativity and motivation of its citizens, and its capability to adapt and respond to unforeseen changes or events.

However, as I argued in Chapter 3, the development of society purely in accordance with the technological imperative would also lead to a totalitarian system. Limiting the criteria for decision-making to those of a committee of bureaucrats and technical experts also reduces the stock of imagination, skill, and experience required for the rational, intelligent, and flexible development of society.[7] The rejection of the capitalist economic imperative and the assimilation of all human activity to the technological imperative is process that would unify art and science in the collective construction of increasing technological power, but the development of increasingly powerful technologies, without any social or humane direction or control, apart from the acquisition of more technological power, would create an unstable, unsustainable, and irrational society because it lacks any vision of human well-being and the good life. The bounds of technical rationality must be evaluated in accordance with such a vision, and, once we recognise that there are a plurality of possibilities and alternatives for such a vision, the very criteria for the evaluation of technical rationality are at stake. The development of the technological society must also be subjected to democratisation – allowing different communities to adopt different directions and levels of technological development – to optimise societal diversity and flexibility. Neither liberal capitalism nor centralised socialism are an adequate ideology for guiding technoscientific implementation and development because their criteria for success will be determined by the interests of a social elite, rather than the needs of the local communities within which the technologies will be implemented and developed. Both of these ideologies lead to the public exclusion from decision-making or only allows the public to participate after all the important decisions have been made. They also tend to reduce the evaluation of proposals to very limited criteria and suppress any critical discussion about alternatives. In both the USSR and the capitalist West, technocratic administration supported the power of the social elite and embodied ideological commitments rather than a practical response to societal problems.[8] Social elites tend to choose to implement the technologies that provide the results that they want and ignore all other effects and their implications for society. Either way, we end up with having technologies and policies imposed upon us by a boardroom or committee. The restriction or limitation of decision-making to specific classes of society, whether an economic or political class, is a societal mistake because it reduces the breadth and latitude of the tactic knowledge, skills, experience, and imagination that is brought to bear in making decisions regarding societal development. This is made even worse when the social elite conspires to limit the access to education for the majority of people, suppresses all forms of genuine political opposition, and actively destabilises the cohesion of communities in order to create a society of disempowered and obedient individuals. This radically reduces the societal capacity for creativity, flexibility, and adaptability by not only reducing the societal stock of tacit knowledge, skills, and imagination of the population, but also disabling the capacity of the population to self-organise at either a local or societal level. Put simply, it is a matter of numbers. Increasing the number of citizens that are involved in the experimental development and organisation of society increases the quantity and diversity of ideas, skills, experience, and creativity that is brought to the task. My argument is that we need to recognise that optimising the capacity for social participation and cooperation is the best way to organise the development of society, but once we realise that pluralism and diversity optimise the flexibility, adaptability, and creativity of our society, then it is evident that the decentralised democratisation of our society is the best political system for people to decide how to develop society. In order to avoid the limits and totalitarianism of centralisation, the economic and political structures of society should be decentralised into local communities governed by the people who live in them. Providing that these communities are able to communicate and interact with other communities, throughout the whole of society, this optimises the creativity of society by maximising participation and embracing social pluralism and maximises the chance for the development of sustainable communities and a healthy society. How a community elects to run its economy and technological development should be a matter of local decision, based on its resources, knowledge, and the ability of its citizens to negotiate and cooperate with their neighbours to obtain what they need or when local decisions are of concern to any neighbouring communities. Once a community focusses on its own development, then there are plenty of opportunities for citizens to look beyond the concept of labour as a commodity to be exchanged for wages, to pay the rent and bills, and, instead, see it as a shared and cooperative social activity to build a better community in accordance with whatever vision of a good community that its citizens come together and decide upon. Even in a culture that values individuality and private property, once the local community becomes the local focus of all economic and political relations, while individual rights and the democratic process are legally protected by the national or federal government, then it is possible for society to optimise the overall lack of any consensus, its pluralism and diversity, by localising and limiting decisions about both means and ends, and enhance the available stock of experience and creativity of society as a whole. It is simply a matter of hedging one's bets by trying all the solutions that people seriously propose. This will not only empower people to experiment and organise themselves when making decisions about the development of their communities, allowing different communities and individuals to try different solutions, as decided by the community members affected by those decisions, but it will also provide society, as a whole, with a greater capacity to adapt to unforeseen events and rapid recovery from disasters. It also facilitates the limitation of the societal impact of any policy or decision to the locality within which it is implemented and developed. It at least acts as a damping factor on the consequences of mistakes. Every day the television and newspapers inform us that we should be thankful and proud that we are living in a democracy. We are informed that we should be so proud of this democratic model that we should export it to the rest of the world as the basis for all governments and societies. Yet, after only a few moments of honest reflection, we are confronted with the contradiction between our pride to be living in a democracy and the fact that we hardly have any real power in influencing, choosing, or changing the formulation, implementation, institution, and development of public policy and the mechanisms of governance, debate, and participation. After such a reflection, we might even find ourselves questioning whether we really live in a democracy at all! So what does it mean to live in a democracy? It is a commonplace view that living in a democracy involves the freedom of every citizen to criticise the government and its policies. However, even though open criticism is a necessary condition for a genuine democracy, it is far from sufficient. Living in a democracy involves the participation of every citizen in the formulation and implementation of government and its policies. Without everyday opportunities for participation in policy formulation, criticism is nothing more than catharsis. It is also essential for democratic participation that citizens are able to develop insight into their social circumstances and the structures that affect their lives. Citizens need to be wary of the uses of propaganda to disseminate ideological justifications for injustice and also to distort peoples' understanding of their circumstances and the obstacles to democratic participation. Ideology can be a powerful source of inspiration that can raise consciousness and liberate people; after all, democracy is itself an ideology, but it can also be used as a means of suppression and deception if it is integrated into everyday life without any critical regard for whether it is true, just, and good. It is crucial for democracy that citizens rationally, critically, and openly examine their beliefs, with an honest concern for truth, justice, and goodness. As Habermas argued, the public life of democratic societies presupposes a commitment by the citizens to engage in rational communication and action. To the extent that we transfer responsibility to representatives, either technical experts or professional politicians, we destroy the very meaning of democracy.[9] The reader may well consider this to be somewhat philosophically idealistic and they would be quite correct to do so, but it is essential for the health of a society that its citizens idealistically engage in a philosophical and critical discussion about their beliefs, norms, and ideals. The most telling aspect of propaganda, in all its forms, is that it is opposed to such a public discussion and the pluralism that emerges from it, given that propaganda is primarily directed to creating social conformity to a unified societal commitment to a course of action (or inaction). We must be most wary of the propaganda of the advocates of "democracy" within the media simulacrum of informed public debate and criticism designed to manipulate public consensus and generate conformity. The task of this kind of propaganda is to leave everyone believing in the rational and progressive nature of the current social order, without having any rational basis for that belief, and serves only the social elite. It is the enemy of democracy. The decentralisation of power to communities acts as the best safeguard against propaganda and the manipulation by mass media. Once decision-making in the development of our communities involves people who know one another, providing that we are secure in the legal protection of our individual and democratic rights, open and interpersonal communication becomes the medium for political deliberation. Once the political process is focussed on local concerns among concerned citizens who know one another, then it becomes increasingly difficult to manipulate those citizens using abstracts and interpretations that have no concrete basis for their lives (even when some of the participants are being dishonest). Also due to the decentralisation of the political agenda, citizens are more likely to be interested in a local media that focusses on local issues, which become important once we can involve ourselves in community development, and, providing that the ownership of media is also decentralised and localised, with local citizens having a democratic control over local media, then a pluralistic media is one the best safeguards against propaganda. The more decentralised and pluralistic the media is, involving many diverse and independent radio stations, television networks, newspapers, and an uncontrolled Internet, then the less likely the propaganda will be successful. If people base their education and process of gathering news, criticism, and information from a variety of sources, without relying upon a unified mass media, then the harder it becomes for the propagandist to mediate between individuals and the rest of society. Once media are decentralised, through community participation backed up by legal rights and protections, then it will become more likely that media will reflect public interests and concerns, leading to more open and critical public participation in the media discussion and presentation of any new ideas and current policies. It will provide opportunities for disadvantaged individuals and groups to increase awareness of their problems, concerns, and ideas.

Strong democracy and public participation

Benjamin Barber defined a strong democracy as being a society that affords the egalitarian participation of citizens in the basic circumstances of their lives and the organisation of society.[10] In a strong democracy, citizens would debate problems and solutions, rather than candidates and parties, setting our own deliberative agenda rather than relying on mass media to do that for us, and, when representatives are needed, the electorate would know them personally. The role of central government would be devolved into legal mechanisms and agencies through which the democratic process could be protected and facilitated. In a strong democracy, it is essential that citizens participate in establishing and preserving the substantive and procedural components of our own processes of decision-making and our own level of participation. It is essential that all citizens can participate in the formulation of the decision-making agendas, but, as Barber pointed out, it would be absurd to demand obsessional participation as being a minimal requirement for strong democracy, in which every citizen is expected to participate with equal zeal in all public debates and decisions. Not only would it be somewhat totalitarian, it would also be impractical and counterproductive. It is not necessary that every citizen actually participates in making every decision, but it is essential that any citizen potentially could participate in making any decision. The minimal requirements for a strong democracy are that all citizens are able to participate in the public matters that are of concern to them, that they are able to set the agendas for debate and decision-making, they have direct access to the processes of deciding policies and laws, and they are able to effectively counter the undemocratic deployment of significant power. Citizens should be able to elect representatives if and when they are unable to participate directly. It is also crucial that the needs, rights, and wishes of socially disadvantaged citizens are also properly represented and protected, when, for whatever reason, they are unable to directly participate or they are suffering local oppression. It would be one of the roles of central government to make sure that all citizens are able to directly participate when they wish, have adequate representation when they are unable to directly participate, and have equal protection under the law. In a genuine democracy, the key function of central government would be to protect the legal rights of every citizen in order to protect their ability to function as citizens.[11]

However, as Barber argued, strong democracy requires a focus of power at the level of local communities because citizens have a much greater ability to participate and exert influence when they have a much more intimate knowledge of the circumstances and people involved. Strong democracies operate on a principle of egalitarian participation, at a grass-roots level, on the part of the citizens in the agenda, form, and evolution of their society. Strong democracy involves a decentralisation of the decision-making process in favour of local communities, organisations, and institutions. Public participation is fundamental for the construction of society and decision-making at every level, and representation plays a supportive and institutional role, rather than as a substitute for participation. Consequently, even though legal and bureaucratic institutions can help and protect ordinary citizens, as we learn how to democratically participate and educate ourselves in how to govern our own affairs, the form and content of democracy cannot be established in advance – it must be dynamically evaluated, chosen, and perpetuated through participation. It is through participation that we will learn how to participate better and ask better questions, transforming democracy into an ongoing social revolution, rather than an institutionalised mechanism for making decisions, and, therefore, it is essential to recognise that an over-reliance on representative, legal, and bureaucratic institutions can stifle genuine democracy and lead to abuses of power by corrupt technocrats, lawyers, and professional politicians.

As Feenberg argued, the real issue is not technology and progress *per se*, but the variety of possible technologies and paths of progress, among which we must choose.[12] From the onset, he adopted the view that modern technology embodies the values of society and especially of its elites, which rest their claims to hegemony on technical mastery. He was critical of the view that the technological society is condemned to authoritarian management, mindless work, and the unrelenting consumption, and rejected the view that technical rationality and humanist values are "contending for the soul of modern man". He considered these views to be clichés. Instead, he took the Marxist line that the degradation of labour, education, and the environment is rooted not in technology *per se*, but in the antidemocratic values that govern technological development, and he called for the development of critical theory to generate a cultural critique of technology, which could rationally evaluate the larger context of technology, articulating, examining, and judging the values and relationships that have become central to the exercise and organisation of political power through technology. Feenberg argued that the political directions of technological development are based on ontological decisions about what it means to be human and what kind of civilisation we wish to construct. He argued that the exclusion of the vast majority from participation in this decision is the underlying cause of many of our problems, and a good society should enlarge the personal freedom of its members while enabling them to participate effectively in a widening range of public activities; therefore, a profound democratic transformation of modern industrial society will resolve these problems. Technological rationality stands at the intersection between ideology and technological power, where the two come together to control human beings and resources in a way that installs the values and interests of the social elite in the very design of rational procedures and machines even before these are assigned a goal, tacitly sedimented in rules and procedures, devices and artefacts, as "technical codes" that routinely reproduces the pursuit of power and advantage. Hence, Feenberg argued that technology should be situated as an ambivalent process between alternatives within a scene of political struggle and debate. Consequently, he postulated that there are at least two different paths of technological development available to us, for example, whether we use computers to control or liberate communication, whether we build our cities around public or private transport, or whether we construct factories as an assembly line or a workers' cooperative, and, hence, there are always at least two possible civilisations that we can choose between when choosing between technologies. He argued that if this choice is made through grass-roots public participation in the implementation of technology, then new paths of technological development and the construction of civilisation will become available to us. However, even though I agree with Feenberg that the ontological development of the technological society depends on societal choices, we need to take a further step back before we can rationally examine and evaluate technologies, in order to choose between them, and examine the rationality of our vision for society, human well-being, and the good life. We need to place these metaphysical conceptions of human being under critical scrutiny. Once we recognise that each technological development is an experiment, then we can take some relief in that fact that, in a pluralistic society, different communities will be developing different technologies in different ways, as well as having different goals and ideals, and we would be able to compare the development of these communities and learn from the ongoing successes and mistakes, but it is essential that we critically develop a democratic evaluation of rationality and the possibilities available to us.

As Richard Sclove argued, the aim of democratic evaluation of technology is not to predict the future, but it is to evaluate the technologies that we already have and whether they enhance or reduce our capacity for democratic participation.[13] The more that we participate in this process, the better we will become at evaluating, reforming, and replacing existing technologies, rather than irrationally innovating new technologies and implementing them for the sake of more power. If this process is performed in an open and cautious way, then we will be able to recognise adverse structural effects quickly, modify and replace them before they become irreversible or disastrous, and make this information widely known to others. Sclove argued that strongly democratic communities should exercise reasonable self-restraint when making decisions in order to avoid "translocal harms", such as pollution, for example.[14] While I agree with Sclove that reasonable self-restraint would help community governance and its relations with other communities, I do not agree that it is prudent for a model of strong democracy to overly rely on this. After all, it seems to me that we would not really be in the predicament of needing to radically reform our political system of decision-making if we were universally capable of exercising reasonable self-restraint, and it raises the question of how we decide what reasonable self-restraint is. What we need to do is raise and answer the constitutional question of how the boundaries of political jurisdiction should shift in order to encompass a broader notion of community when local decisions are perceived to potentially lead to "translocal harms". What we need is an inclusive concept of community enshrined in constitutional law. When the consequences of local decisions are likely to extend beyond the local community, then a broader notion of democratic community should be constitutionally available in order to facilitate the participation and representation between members of all involved and concerned citizens. It is an implication of such a notion that the boundaries of "the community" are capable of being shifted to include all those concerned with the wider consequences of local decisions and possible "translocal harms". The boundaries of political jurisdiction of the democratic process should be flexible enough to adjust in accordance with the perceived need of wider participation of those concerned with any local decision and its potentially regional or national consequences. A strong democracy requires a broader notion of community than that of geographical proximity – it requires a network of relations and interests among people who know and communicate with each other. It requires shared lifeworld experiences and projects. Thus any individual may well be a member of several communities, not limited to geographical location. If the democratic process constitutionally demands that shifting boundaries of jurisdiction are required in order to include all concerned citizens in any decision, then there is no need to rely on reasonable self-restraint. The exclusion of concerned and involved citizens from the decision-making process would make any resulting decision unconstitutional, undemocratic, and illegal. It is at that point that excluded citizens could call on the intervention of the regional, national, or federal government. In a strong democracy, the function of government would be to protect and empower the democratic process within local communities, but the mediation of the democratic process must be left to the local citizens of those communities. The exact nature of relationship between local communities and central government, whether it is comprised of a stratified or federalised system of democratic forums and institutions, how participation and representation are to be combined and balanced, involve complex questions that are, unfortunately, beyond the scope of this book. However, the minimum requirements of the relationship must be that the processes of governance are transparent, all citizens have equal rights and representation, and citizens have access to and can participate in government at all levels. It is also important that citizens from different communities also can join together in NGOs that can facilitate collective action and, hence, act as a balance against abuses of power at both governmental and local levels. These NGOs can also help the democratic process and facilitate the exchange of knowledge, skills, ideas, and resources between different communities. It is important for the health and sustainability of a society that the search for commonality does not degenerate into social intolerance, conformity, bigotry, and xenophobia. Intercommunity exchanges and affirmation of the societal value of diversity and pluralism are essential for the health of society; therefore, as Sclove argued, even when consensus is hard to achieve, a genuine commitment to strong democracy increases the level of mutual respect and tolerance for other people.[15] When all participants feel respected by the community, then there is a greater chance of achieving mutually beneficial agreements, actually listening to the ideas of other people, and also citizens are more likely to accept decisions that they disagree with.

One of the criticisms of strong democracy is that nothing would ever get done if everyone was involved in deciding what was to be done. However, this is not really a good objection. It not only presumes that the qualification for strong democracy is that everybody decides everything, but it also presumes that it is always better to do something quickly rather than wait until there is a well-thought-out course of action. Of course, genuinely democratic decision-making will take much longer than authoritarian dictatorship. Perhaps that is unavoidable, but it is also clearly the case that any decision is more likely to be a well-thought-out decision if it is made and agreed upon by all the people likely to suffer the consequences of that decision. It may well take a long time for a community to come together and agree on important decisions, but the reward of this effort is that the decisions that it makes are likely to suit people for a long time and perhaps, with luck, only need modifications or further elaborations to adapt to changes in circumstances. This is actually much more efficient and sustainable than having to follow the arbitrary dictates of a tyrant, even if he or she can make them in the blink of an eye, because arbitrary dictates will probably need to be completely different tomorrow or they will simply fail to achieve their goal. This is even more evident when we consider the historical cases of all the squanderous and murderous efforts made by tyrants and corrupt governments to cover up their bad decisions and suppress all criticism. By making participation open and accessible to all those that wish to participate, on a voluntary basis, it is also evident that a great deal of action can occur without any widespread interest or involvement. Democracy does not imply that an elected committee or ever citizen must debate and ratify every decision (as if the decision about whether to build a nuclear reactor to supply a region with electricity has the same social weight as the question of whether I should have mashed potatoes or yams for my dinner). All that strong democracy minimally requires is that there are social processes, through which people can bring their concerns to public attention and deliberation, socially initiate change in policy or practices, and organise participation in setting the agenda and a plan of action. As long as there is an enforced legal framework that protects the democratic and individual rights of citizens, allowing all those concerned to participate in decision-making, as well as possibility for communication and shared information, people will be able to learn about proposals and participate, if we so wish, in the decision about whether and how any proposal should be implemented and developed. This legal framework would allow citizens to be able to shape legislative and electoral agendas and procedures, when we are affected by those agendas and procedures, in order to develop and adapt them to the democratic needs of different communities and our different circumstances. It is important that the democratic process itself can be adapted to the plurality and diversity of local needs, concerns, experience, and ideals, and, providing that citizens are protected from abuse and disenfranchisement, this degree of flexibility is itself highly desirable for the development of strongly democratic communities. It is important to realise that the public participation in the design and development of communities is not simply a matter of electing to proceed with one design proposal over another. Public education about how to participate in the development of urban areas is not just about developing the infrastructure of our communities, but about developing democratic participation.[16] The immediate benefit achieved from public participation in the whole design process is the improved ability to participate through the acquisition of experience and skills about how to participate. The more ordinary citizens are involved in the design of our communities then the more intimate we will become about the processes of how decisions are made within a complex society and how social structures have been constructed to prevent and hinder public participation. If the democratisation of community development is to become a reality, then it is imperative that we learn and develop strategies for diverting economic and political resources and powers into our communities. As well as putting pressure on the current political order to allow greater public involvement and representation, this involves actively participating in the development of our communities, using our current knowledge, skills, and resources, without waiting for the permission to do so. This involves deciding how to use the available technologies and resources through democratic forums and processes that we democratically agree upon. It also involves teaching each other basic skills that allow us to become more self-sufficient. It involves collectively boycotting developments, businesses, and industries that are imposed upon our communities, unless we are able to participate in whether and how they are to be designed, implemented, and developed. Of course, this will result in difficulties, inconveniences, struggles, and failures, but if we wish to live in a democracy, then we must come together and take control of how our communities are developed. This requires great efforts on our part. Otherwise the decision-making process and power will remain in the hands of a social elite that is relentless and tireless in its efforts to come together and take control of how our communities are developed.

Another objection against strong democracy is that it would inevitably involve protectionism and the regulation of regional, national, and international trade, and, as a consequence, it is too idealistic to work in the real world of globalised trade and large corporations. This objection is that the protection of local economies runs against the international economic system (institutionalised through the General Agreement on Tariffs and Trade, the World Trade Organisation, the World Bank, as well as other international organisations committed to globalisation) and would effectively disband that economic system in favour of decentralised local trade agreements and an international commitment to provide sufficient democratic representation and participation to effectively democratise world trade; the organisation and protection of labour and workers' rights; egalitarian access to communications, technology, and transportation; the protection of indigenous peoples and environments; the dissemination of technoscientific knowledge and skills; and the establishment of pluralistic democratic forums for the representatives from every nation to discuss ideas, concerns, and grievances (without being dominated by the most powerful and wealthy nations). It is claimed that this would be impractical or impossible. My response to this objection is a straightforward question. If it is impossible to democratise world trade and politics because of the established vested interests of powerful nations and corporations, then to what extent can we describe Western Civilisation as being an advocate of global democracy or comprised of genuinely democratic countries? It is also frequently argued that any form of workers' rights increases the chance of capital flight to countries that do not provide workers' rights. However, the more economically diverse and self-reliant that a community becomes, then the more it is able to accommodate the consequences of capital flight because the community does not depend on a single industry and is thereby able to make demands about the working conditions and rights of its citizens. Hence, one of the important democratic functions of government is to protect the right of every community to govern its own economic development by legally protecting and encouraging sustainable economic relations between communities that are on an economically equal footing, as far as levels of wages and workers' rights are concerned, hence promoting fairer trade and protecting local economies, and also offering more opportunities and incentives for local citizens to start their own businesses and workshops to satisfy local needs. This would prevent large corporations from taking economic advantage of their antidemocratic policies (such as using coerced cheap labour in underdeveloped countries to control markets in developed countries). Of course, in order to safeguard sustainable economic relations between communities from abuse and corruption, it would be necessary to establish regional, national, and international procedures for monitoring trade, democratically addressing grievances, and punishing offenders. Once communities are protected by preventing exploitative and antidemocratic trade, large businesses and corporations would be economically compelled to continue to invest in developed and democratic communities, even with their higher labour costs and levels of workers' protection, because if investors wish to take advantage of the markets of those communities, then they have to invest in local commerce, industry, and agriculture. Moreover, the localisation of trade allows citizens to know all the participants, where the goods come from, and how they were produced, which develops a rich lifeworld relation with our material conditions.[17] It is on the basis of our experiences living in a community that we learn how to nurture collective practices, mutual respect, and develop a sense of commonality and friendship with our neighbours. These are basic experiences for the development of democratic organisations and institutions that are politically based to establish collaborative relations, such as economic relations, with people from other communities and countries. It is this ability that is essential if we are to achieve enduring international peace, justice, environmental protection and sustainability, and mutually beneficial trade agreements.

Sclove noted that a study of several thousand American workers concluded that they spend more mental effort and resourcefulness in getting to work than in doing their jobs.[18] The physical organisation of the lifeworld, especially public spaces and the working environment, reflects the power relations of society and can act as an obstacle to democratic participation. It is important that we democratically participate in the design and construction of the public spaces and working environments of our communities. The power relations inbuilt into the organisation of labour and the working environments within which most of us spend much of our adult lives naturalise structures and hierarchies of dominance and conformity within their technological infrastructure and operational procedures. It is essential that we are able to transform these power relations through democratic participation in the organisation of our work environment, whether it is in agriculture, industry, or commerce, and this can only be achieved if we are able to democratically participate in the formulation of strategy, management, and policy in accordance with our needs and the needs of our communities. Of course, investors will want to receive a good return on their invested monies, but that does not justify the subordination and abuse of workers and our communities. It is also quite reasonable for technologists to aim at the maximisation of efficiency and productivity, but that does not justify the imposition of technological innovations on communities and all the changes that such experiments inevitably induce. In a genuine democracy, the capitalist economic and the technological imperatives must be subordinated to the public good, as decided by the public. It may well be the case that particular workers and communities accept and agree that the capitalist economic imperative and/or the technological imperatives are in the public good, for those workers and communities, but that is a matter for local democratic consensus, rather than being treated as if they were *a priori* truths that can never be rationally questioned and criticised. The democratic process requires such a rational process of questioning and criticising the ends and means of the economic, political, and technological development of our communities. We should structure our own labour in accordance with our perception and consensus about the local needs of our communities and how we wish to balance the time spend on work, leisure, and political participation. This will allow more citizens to participate more in the political process and that will help to create greater social equality and responsibility. Sclove suggested that work-release time to enable citizens to have the time to perform political participation could well be established in a way that is analogous to jury service.[19] However citizens choose to do this, either through job sharing on an individual basis or careful organisation of the whole community, through increased childcare support, through permitting flexibility about the retirement age, through encouraging better paid part-time employment over poorly paid full-time employment by improving the efficiency and productivity of businesses, farms, and workshops, or more imaginative ideas than these, are matters of local decision and experimentation. Unless people are able to live and work convivially, creatively, and with a level of responsibility that challenges our abilities and aspirations, as well as equally sharing social burdens and tiresome tasks, we will be unable to truly achieve the societal conditions that are necessary for egalitarian and participatory democracy. Democratic changes in how work is performed will only occur if the labour process is rationalised in order to optimise the intrinsic creativity and importance of the work for each and every worker, as well as the level of participation in the decisions regarding the tactical and strategic management of the work. Of course, there needs to be a balance between maximising the aesthetics and productivity of the labour process, but boring and dangerous labour should be automated, whenever possible, and shared, when automation is not possible. Advances in the technological processes of production should be used to shorten the average working day (rather than reduce labour costs) and permit the worker more time for leisure, education, family, and political participation. As Marx argued, increases in productivity should be used for the benefit of workers rather than just to increase the wealth of a social elite. We should be able to enjoy our working life as much as possible, after our labours have satisfied our material needs and the needs of our community within which we live, and, therefore, we should be able to have considerable plurality and latitude in our choice of labour. Democratic participation in factories, workshops, offices, farms, as well as other organisations and institutions, would inevitably increase the labour and running costs (lowering the profitability), but because it would also inevitability increase pride and a sense of community, it would reduce the incidents and costs of illness, sabotage, depression, alcoholism, violence, drug addiction, crime, strikes, and unemployment. Hence, while it would reduce the dividends of shareholders, it would also reduce the tax bill too because it would require less social services to deal with the problems caused by overwork and work-related stress. Moreover, through democratic participation, the costs of local governance would be reduced as the need for a bureaucracy is decreased, and its citizens share the social burdens of the community, become increasingly aware of the needs of the communities within which they live, and develop a greater stock of skills and resources for the satisfaction of community needs. Why should a community pay taxes to a bureaucracy to tender an irrigation project to an outside contractor, when after a few public meetings, the citizens of that community could easily come together to design and build one for themselves for a fraction of the cost? By developing a plurality of skills and cooperating with one another to improve the community's infrastructure, citizens would be able to enjoy higher wages and lower taxes, a much higher quality of life, and live in an increasingly sustainable and increasingly self-reliant way. Moreover, citizens would be able to decide the local price of the community produce, and, as a consequence, higher wages would not necessarily lead to inflation, as if it were a natural law that one follows the other.

Once a community's local economy and public purse is under the governance of its citizens, then it becomes possible for local people to collectively invest in technological projects that we agree have social and political benefits for our communities, in accordance with our vision for the development of our communities. However local democratic procedures and forums are decided by local citizens, it is essential that public participation is not limited to the evaluation and ratification of contracting projects, but is also involved in the overall implementation, development, operation, management, and monitoring of the technological infrastructure of that community. Through participating in the everyday operations of a wide range of technological infrastructures and practical activities, we increase our technical knowledge, skills, and appreciation of the wider implications of implementing and developing technologies. Moreover, we would gain increased social skills and critical appreciation of difficulties, implications, and responsibilities of democratic participation. Local people who know and interact with each other should satisfy local needs, as much as is practically possible, because this keeps power and resources in the community and also develops a sense of self-reliance, communality, and social responsibility. It is better if we are able to build and repair the technological infrastructure ourselves, using the skills, knowledge, and resources of our own communities, but when outside help is needed it is essential that local people have overall control of the project. Of course, as Sclove pointed out, apart from in a few Amish communities, we lack an example of a societal tradition of subjecting technologies to democratic scrutiny, evaluation, and control.[20] Our tradition is that bureaucrats, politicians, and corporations make these decisions for us, which, of course, they represent as the technically rational decision in accordance with technical and economic factors determined by impersonal natural laws and market forces, while they make the decision which best suits their interests. The technically rational decision is represented as progressive, even if it remains untested and experimental, and is often used to oppress local communities that have traditional, sustainable, and well-tested technological practices that are sufficient for their local needs. Hence, it is often argued that democratic participation would slow down the process of designing and constructing technological projects. However, this is not an intelligent objection. It is much better to explore every conceivable aspect of a proposed project before it even reaches the drawing board, rather than having to deal with the expensive and disastrous consequences of an inappropriate or ill-conceived project after it has been constructed. It is much better to take advantage of local knowledge and participation during the planning process than to ignore or suppress it. It is quite simply more cost effective to design and build a sustainable, locally appropriate technology, and it requires local knowledge and participation in order to have a better chance of actually doing that.[21] While there are many philosophical, political, and practical problems that would need to be addressed in order to optimise the chances of developing strong democracy and public participation, all of which are beyond the scope of this book, such as the relations between the communities, the constitutional structure of participatory democracy and the role of representation, democratic rights and obligations, commercial and military secrecy, media and propaganda, and many other such problems, the rest of this book shall be concerned with the claim that the process of technological evolution is just too complex for ordinary citizens to make the correct choices about which technologies should be developed and implemented because ordinary citizens are simply not competent to participate in the nuts and bolts of the design process. However, as I shall argue in the last section below, even though this objection makes an important point about the limitations of public participation in the design of technologies, it presumes that design choices and specifications are defined by the internal workings and technical operations of the technology, in relation with their intended results, whereas the democratisation of the technological society involves a much more holistic and pluralistic view and does not involve the removal of the input of technicians, economists, bureaucrats, and professionals. It is simply the case that the decision-making agenda should not be pre-emptively constrained to technical choices and judgements. There are many different ways that the public can be involved in the design of technologies, at every level, and the greater the degree of participation that ordinary citizens have, then the more they will appreciate the technical problems and limitations involved in making choices and judgements when designing technologies. Furthermore, as Sclove pointed out, there are numerous cases of technological projects that successfully involved public participation, which failed only due to outright opposition from powerful institutions.[22] The ability of ordinary citizens to participate in R&D is not so much a question of technical competence, but, instead, is a question of how citizens can overcome bureaucratic, economic, and political resistance to public participation. The most difficult challenge for strong democracy is not how to decide which technologies to implement and develop, but how to overcome the power of antidemocratic elites that see public participation as being a threat to their control and authority.[23]

Democratising the technological society

How can the technological society be democratically developed? Winner argued that modern technological and political systems tend to promote centrally co-ordinated, technocratic administration and hierarchies that use "technical experts" as an instrument for the preservation of the *status quo*.[24] As Sclove pointed out, much of the prevailing technology and architecture of our society are designed and deployed in accordance with authoritarian purposes. They function materially and ideologically against democratisation, and, therefore, we should not expect an effortless transition between authoritarianism to democracy.[25] As long as "the technical" and "the social" are represented as being distinct aspects of modern society, then the technological society will remain incompatible with democracy. It is essential to critically examine this distinction in order to show how "the technical" has been placed over and above "the social" in the political decision-making process, in order to limit the criteria through which any technological processes are publicly evaluated – often limited to costs and risks – and established prior to any public consultation, to direct and limit democratic participation in the technoscientific development of society. However, as Winner argued, examining technological development solely on the basis of narrow evaluations of costs and risks ignores the larger moral and political dimensions. If citizens are going to have a genuinely democratic role in the construction of the technological society, then we must be able to participate in deciding societal goals, setting the agenda, strategic planning, design, education, communication, public contracting, policy formulation, media deliberation and criticism, seeking alternatives, and the actual processes of the implementation and development. As Sclove argued, strong democracy provides: opportunities to discover the social contingency of any technological endeavour, which would otherwise be obscured; motivation for the public development of participatory competence in technological politics; a broadening of critical reflection on public needs, ideals, and concerns; a deeper examination of the possible consequences of any new technology; and an increased awareness of the antidemocratic structuration of our technological infrastructure.[26] Early public participation in the research and design of any prospective technology allows for accommodative adaptations throughout the whole process, which leads to the implementation and development of more flexible and socially responsible technologies, greater public receptivity to new R&D ideas (rather than suspicion that yet another technology is going to be hurriedly imposed on an unwilling public), greater likelihood of general satisfaction with results, and a more fair and inclusive process of implementation and development. The democratisation of the technological society requires the development of democratic procedures for the evaluation of technoscience; democratic participation in the implementation and development of technoscience; and the development and implementation of technologies to help the democratic process.

However, a crucial problem for the democratisation in a technological society occurs when "technical experts" are required in order to understand technical details required to make informed choices between possible directions of technological innovation and scientific research.

How can citizens participate in scientific and technical decisions? How can "the social" inform "the technical" in these contexts? How can we rationally democratise the technological society? As Raphael Sassower pointed out, even though "technical experts" should act as translators between the public and the technoscientific community, the public needs to be aware that such translators have their own agenda, overly simplify the situation in order to be able to communicate with the public, conceal problems in order to gain public acceptance for a proposed project or calm public fears, and misrepresent their degree of certainty because of the perception that the public requires certainty.[27] Sassower argued for an examination of our expectations regarding the feedback loop between the public, the "technical expert", and the technoscientific community. He was concerned that the technoscientific enterprise would be damaged if it caters to the anxieties of the public and "constantly pandered to the public and its expectations" for certainty and assurances because it would be disastrous for the public if the technoscientific community was forced to conceal its errors, failures, and doubts. Marx Wartofsky argued that this is a political problem concerning the process through which the norms and definition of rationality are to be achieved.[28] For Wartofsky, the democratisation of technology requires the technoscientific education of the public, alongside with the emergence of a public awareness of the undemocratic distribution of knowledge and power within the construction of the technological society. Rational democratisation of technology requires increased access to technoscientific training, as well as a critical, political engagement with the inequalities of access to technological power. While I completely agree with Wartofsky's claims about the need to politically address inequalities in the distribution of technological power within society, requiring greater levels of technoscientific education for the public, in my view, it is also an essential requirement for the democratisation of the technological society that scientists and technologists are also educated in alternative forms of rationality and the wider criteria involved in the development of the quality of life. It is not just the case that we need to become better technoscientists, so to be able to rationally participate in the decision-making process, but also technoscientists need to become aware of the non-technological criteria involved in rationally evaluating human goods and societal progress.[29] Engineers need to be well read in the humanities, while, at the same time, dedicated to promoting a high level of humanitarian culture in society. Technoscience cannot be developed in isolation in accordance with the centralised plans of administrative bureaucracies and social elites. The development of the technosciences requires broad public participation and initiative as its basis. In other words, it is essential that technoscientists become better members of the lay public and that we dissolve the equation between technological and political power. Thus, the political power over technological power must be developed in accordance with a pluralistic debate about the norms and definition of rationality over and above bounded technical rationality. It is the task of coming together and deciding how this debate is to be conducted that is the fundamental realisation of the potential for democracy in a technological society. It is not only impossible to adequately define the framework for this debate from the onset, due to the pluralism and complexity inherent and emergent from ongoing critical debate, but it is actually counter-productive because it would be antidemocratic to do so. Thus the democratic debate about technoscience is not necessarily just about how we effectively and rationally control and shape the development of technoscience and society, but it is also going to include a critical debate about the overall goals, ideals, and values of democracy itself.

Of course, ordinary citizens would benefit from an increased access to technoscientific education and literacy. This would help citizens make informed choices. However, the most important criteria for the decision-making process (such as the knowledge of how to achieve the human good life and universal happiness) are those for which there is an absence of "technical expertise" or even any universal consensus, for that matter. Once we recognise that the rational development of technoscience is bound together with the rational evaluation of human well-being, the good life, and the ideal society, then pluralism, diversity, and democratic participation need to be recognised as being of paramount importance for technoscientific research and development. As Sclove pointed out,

> However, if the most important knowledge about a technology involves not its internal principles of operation but its structural bearing on democracy, then presumably the latter kind of knowledge should constitute the very core of technological literacy. Yet experts, even the elite, typically know little about this first-order issue – not even that it is an issue. Must one not reluctantly include among the technologically illiterate – in that term's socially most meaningful sense – the majority of technical experts?[30]

Moreover, once we recognise that the consequences of the implementation and development of technologies in the world cannot be determined in advance and predicted, then we should no longer accept that idea that the "technical experts" have any greater capacity to make decisions about the technological development of society than any other members of the public. Once we understand that every techno-logical innovation is transformed once it begins to interact with other technological systems, as well as other complex features of the world, we must also understand that "technical experts", as specialists in a narrow technical area, cannot claim to have any expertise regarding the societal and environmental consequences of new technologies. Most "technical experts" are completely technologically illiterate about areas outside their specialisation. As the complexity of technology increases and its non-linear interactions within the world become increasingly ambiguous, "technical experts" become increasingly unable to demon-strate any certainty or foresight over and above that of the ordinary citizen. In this regard, the "technical experts" are just as much members of the public as the rest of us. This is also true of professional politicians, "political experts", and bureaucrats, who are no more competent (or less incompetent) than any other citizen.[31] If both "technical experts" and professional politicians are on a par with ordinary citizens when it comes to the question of the social and environmental consequences of new technologies, then it is quite foolish to rely on "technical experts" and professional politicians to make decisions for us. They will simply make guesses and decisions that suit their own agenda and interests. Unless they can demonstrate some objective and rational knowledge about what is good for humanity as a whole, we would be wise to presume that there are not experts on that question. It was for this reason that Larry Hickman argued that the crucial development of democratic parti-cipation in the technological society does not depend on making each member of the public into a "technical expert", but, instead, depends on reforming and maintaining the potential for enlightened debate concerning the vital interests of the public, free from manipulation by mass media propaganda.[32] Thus, the task of public education is not simply that of imparting facts, knowledge, and skills required for the acquisition of "technical expertise", but it is about awakening effective intelligence in public affairs and democratic participation. Furthermore, as Sclove pointed out, if the complexity of technology is inherently at odds with the practical possibility of developing a strong democracy, then, perhaps, it is our obsession with technological innovation that we should be questioning, rather than our democratic ideals, and we should be very wary of leaving technological implementation and development in the hands of "technical experts" and professional politicians. As he put it,

After all, it was not panels of laypeople who designed the Three Mile Island and Chernobyl nuclear plants; who created the conditions culminating in tragedy at Union Carbide's Bhopal, India, pesticide factory; who bear responsibility for the explosion of the U.S. space shuttle *Challenger*; or who enabled the *Exxon Valdez* oil spill.[33]

It may well be technologically imprudent – as well as undemocratic and complacent – to leave technological decisions in the hands of the "technical experts" and professional politicians. The implementation and development of technology should be subordinate to the democratic process, and ordinary citizens should not allow the "technical experts" and professional politicians to dominate the agenda regarding technological policy. The contingency and value-ladeness of technological choices should never be dominated and monopolised by supposedly impartial experts. Experts have a vested interest in arguing that the most important questions are those that the expert can answer. They tend to view technology as a means to solve the specific problems that it is posed as a solution for and they tend to limit their analysis to quantifiable criteria, such as construction time, job creation, revenue, cost, risk, investment return, and durability. However, there are many other criteria that need to be considered, such as long-term environmental and social impacts, transformation of human character and the lifeworld, and also moral and religious concerns. It is of paramount importance for the health of society that these qualitative and complicated criteria shape and limit the bounds of technically rational decisions. The belief that all technological change is progressive and beneficial is naïve, and it is highly arrogant to impose new technologies on a community, especially given that every technological change is a social experiment. It is not only fair that communities should be allowed to consider whether they wish to be experimented upon, but there is also a better chance of success for any experimental introduction of new technologies into a community, if the community participates in the implementation and development of those new technologies. All decisions are made in the absence of any certainty regarding the social and environmental consequences of any technical action, whether constructed in accordance with a narrow technical agenda or a broad democratic one, and, consequently, the best reason for democratic participation is that broader criteria increase the chances of anticipating problems and suggesting remedies or alternatives. The complexity of technology is not a good reason to exclude democratic participation. It is a good reason to encourage it.

Mumford advocated an "organic ideology" for society that called for the rational and harmonious convergence of the practical concerns of the world and the need for fostering sustainable relations between all aspects of life.[34] Once we think about life in its full manifestation in terms of an organic whole, the continuing refinement and integration of technology, subordinate to the development of harmonious relations with all aspects of life, promises greater opportunities for human flourishing to become universal and rationally developed. I agree with Mumford that the challenge facing us – something which is even more imperative today than during the time of Mumford's writing – is how to use science and technology (rather than avoid them) to transform the technoscientific development of modern society in such a way as to place it at the service of bettering life for all, rather than the privileged few, and creating a more sustainable form of the lifeworld that optimises the chance for as many people as possible to live a good life. The pernicious and damaging effects of technology are a consequence of our failure to integrate technology within a rational and cooperative society. However, argued Mumford, a mature society, as part of the maturation process, must come to understand the rational limits of our technical abilities that are imposed "by the very nature of the elements with which we work."[35] It is through such an understanding, once we have developed into a rational and cooperative society, that we will be able to develop states of sustainable equilibrium. We need to move beyond the confrontation with Nature and the naïve faith in our ability to construct a better artificial world. Instead, we should work at integrating our technologies within the natural world, in order to choose and develop sustainable and rational ways of life, and build our vision of a good life into the sustainable development of our communities. As I have argued in this book, human beings have developed a society as a confrontation with Nature, within which the natural world is appropriated as simply being instrumentally available, or is represented as something dreadful and capricious which can destroy all our efforts without warning. This attitude of confrontation is brought very much into question when human activities, such as deforestation or pollution, either cause environmental disasters or intensify natural events into disasters. We should learn to recognise the extent that how we participate in the world irreversibly changes the world in ways that we simply cannot foresee. The challenge facing us today is how we can develop our technoscience in order to live harmoniously within the natural world. As Mumford argued, this will involve the development of environmentally sustainable methods of constructing and organising cities, transportation, industry, agriculture, and mining, as well as conserving the natural world. This will involve a greater depth of understanding of the possibilities and benefits of cooperating and harmonising with natural processes rather than competing with or replacing them. When practised within a democratic society, it will help prepare, forewarn, and adapt us to the capriciousness of Nature, as well as our own limitations, by optimising our capacity to adapt and change in response to changes in the world.

As Thomas Hughes has argued, following on from Mumford's call for an integration of technics with the organic, we need to integrate our concerns about the conservation of the natural world, the preservation of the health, integrity, and diversity of the environment, within the human-built world, in order to integrate ecological science and modern technology in order to construct an *ecotechnological environment*.[36] When architects take into account the local climate and landscape when designing building, then they are creating an ecotechnological building; when engineers and planners take into account the natural flow of rivers, streams, and prevailing winds when designing irrigation systems, then they are constructing ecotechnological systems. Hughes argued that this movement found its roots in the American puritans' idea of creating a pastoral New Jerusalem in the new world. Citing Perry Miller, Thomas Jefferson, Robert Beverly, J.A. Etzler, and Ralph Waldo Emerson, Hughes described how the early American settlers proposed a poetic vision of using science and technology to confront the wilderness and create pastoral communities that would nurture genuine virtue and return human beings to a state of grace.[37] Through the prudent and careful construction of mills, canals, and steam engines, human beings could build a new country of interconnected, but self-sufficient communities, within green valleys, filled with orchards, farms, and the sound of song-birds. Agricultural combines would reap, sow, and plough, converting the wilderness into a bountiful garden landscape, free of the squalor and satanic mills of Europe, to bring about a new Eden within which all human beings would be equal, free, and happy. However, as history testifies, things did not quite work out that way. As Hughes put it,

> Technologically empowered, we have reason to doubt our values and competence as creators of the human-built world and as stewards of the remaining natural world. Slums in inner cities, ugly strip malls in the suburbs, hastily and cheaply built housing, polluted air and water, the loss of ecologically nurturing regions, and the likely threat of global warming give evidence of our failure to take responsibility

for creating and maintaining aesthetically pleasing and ecologically sustainable environments.[38]

Today's urban planners and developers often use technologies that ignore or replace natural terrain rather than interact and adapt to it. This can destroy natural areas that previously acted like natural buffers and defences against flooding, wind, erosion, and so on. Thoughtlessly destroying these areas increases the chance and intensity of natural disasters. Hughes argued that the degradation of the natural world and the human spirit has been of our own making, and, therefore, we should not blame God or gods for the natural disasters that are caused and intensified as a result. It is all too easy for us to blame someone else, such as industrialists, developers, architects, engineers, bureaucrats, scientists, and politicians, but we want cheap industrial and agricultural products, our own cars, low taxes, and we welcome and elect those that promise these things. There are also numerous ways that we can participate in the decision-making process, or at least make our protest regarding our exclusion, but most of us decline to do this and leave it to others. He argued that we should take responsibility and participate more in the design, construction, and development of the human-built world. In order to do this, claimed Hughes, we need to become more technically literate and informed, and then we can consciously and purposefully use technology to shape our world into an ecotechnological world.

The argument of this book has been that it is essential that the ontological construction of the technological society is developed democratically in order to embrace the plurality of alternative conceptions of rationality, human well-being, the good life, and the ideal society. Otherwise a community could make disastrous and irreversible decisions that undermine its own sustainability and severely damage the natural world within which that community (and others) are situated. The challenge facing us is to democratise the construction of the technological society – otherwise we should not be surprised that our society embodies the short-term and narrow interests of a social elite and their technocratic employees. The problem of negotiating between our short-term and long-term interests is a crucial problem for every community. However, providing that the democratic process is maximally inclusive and there are clear benefits to the long-term investment in communities, it will always be the rational choice to act in ways that promote the long-term sustainability of a community rather than short-term gains without regard for the future. Ecotechnological designs involve the development and implementation of technologies that harmonise with the social relations and structures of our communities as well, as well as those that harmonise with the natural world. They must be carefully integrated in relation with both the social and natural environment, which involves maximising public inclusion and local knowledge in the design process from the onset, as well as taking the advice of ecologists, other scientists, and technical experts into account. By maximising public inclusion and local knowledge, as well as technoscientific knowledge, the design process will increase the chances of developing a sustainable technological infrastructure for our communities, which embraces the broadest criteria possible for evaluating the public good. Such ecotechnological planning will lessen the human impact in the natural world, as well as lessening the impact and intensity of natural events by making us less susceptible to their potentially disastrous consequences. I also think that it would lead to better designed cities and a greatly improved quality of life for large numbers of people. If our cities were built on principles that respected natural features in order to adapt human purposes to the natural terrain, then this would create an ecotechnological environment within which fresh air, clean water, parks, and the sustainable use of resources would be commonplace. An increased access to technoscientific education and knowledge will help the general public critically evaluate proposed technologies and scientific research, but it is not the most important problem facing the ability of the public to participate in the decision-making process. The biggest problem occurs when the increased complexity of technology is rhetorically used to justify limiting decision-making to bureaucrats, technoscientists, and professional politicians. If the public is involved, it is usually after all the crucial decisions have been made. But, as I have argued above, technical experts, bureaucrats, and professional politicians are not particularly better placed than any other member of the general public to make decisions based on the societal merits and dangers of any proposed technology. Most of the possible points of interaction between a new technology and the aspects of the real world within which it is implemented will involve aspects and features that cannot be accurately predicted. Nobody can foresee what all the serious consequences of implementing new technologies will be and how those technologies will be transformed by and transform the aspects of the wider world within which they interact. Each decision is an experiment made on the basis of bounded technically rational judgements, at best, or without any consideration of the long-term consequences, at worst. The complexity of technology is not a good reason to exclude the public from the decision-making process because, on that basis, we would have to exclude the politicians, bureaucrats, and the technoscientists as well. In fact, once we accept the ontological complexity of technology, then it is clear that wide public participation is advantageous for broadening the scope and bounds of the processes through which we can decide what the evaluative criteria should be. Even though there are no guarantees, the most important aspect of public participation is about deciding how to rationally and critically evaluate what the good life is, how we should build our communities, and how we should envision the overall structure, ideals, and practices of the whole democratic process. Everyone is on a par in that regard.

Notes and References

1 The capriciousness of nature

1. See Drees, W.B., *Religion, Science, and Naturalism* (Cambridge University Press, 1996); and, Brooke, J.H., *Science and Religion: Some Historical Perspectives* (Cambridge University Press, 1991) for interesting, contemporary theological discussions of this point.
2. See also Hilton, H.B., *The Age of Atonement: The Influence of Evangelicalism on Social and Economic Thought 1785–1865* (Oxford University Press, 1988); and, Buckley, M., *At the Origins of Modern Atheism* (New Haven: Yale University Press, 1987) for interesting and very different discussions of this point.
3. *On Nature, the Utility of Religion, and Theism* (Longmans, Green, Reader, & Dyer, 1874).
4. Charles Darwin, for example, was appalled by the apparent squanderous cruelty and brutality of Nature. He found the idea of the natural world being the intentional creation of an all powerful and good god in accordance with a divine plan to be quite an unbelievable idea. The egg-laying habits of the Ichneumonidae (which lays them into a caterpillar host), disease (which killed his ten-year-old daughter Annie), and earthquakes were considered by Darwin to be serious evils in Nature (and contradictions against the idea of a beneficent deity). See Desmond, A. and Moore, J., *Darwin* (London: Michael Joseph, 1991) and Brooke, J.H., "Darwin" (in Brooke and Richardson, eds., *The Crisis of Evolution*, Milton Keynes: Open University Press, 1974) for further discussion and references. A similar posture of moral outrage against the blind cruelty and brutality of Nature can be found in Williams, G.C., *Plan and Purpose in Nature* (London: Orion Books, 1997) and Rolston, H., *Environmental Ethics* (Philadelphia: Temple University Press, 1988), but in these cases this moral outrage is used to justify a Christian and Kantian account of morality as being opposed to instinct.
5. For example, see Raup, D.M. and Sepkoski, J.J., "Mass Extinctions in the Marine Fossil Record", *Science*, 215, 1982, pp. 1501–3; Ward, P.D., *On Methuselah's Trail: Living Fossils and the Great Extinctions* (New York: Freeman, 1993); Myers, M., "Mass Extinction and Evolution", *Science*, 278, 1997, pp. 597–8; Nee, S. and May, R. M., "Extinction and the Loss of Evolutionary History", *Science*, 278, 1997, pp. 692–4; Hoffman, P.F. and Schrag, D.P., "Snowball Earth", *Scientific American*, 282, 2000, pp. 68–75.
6. For a contemporary scientific focus upon the apparent indifference of Nature towards individuals, see Dawkins, R., *The Blind Watchmaker* (Harlow: Longman, 1986).
7. Of course, it may well be the case that the cosmic order may well destroy human beings (or some other dominant species, from time to time) in order to restore balance – as in Sophocles' *Antigone* (New York: Dover Publications, 1993) – while being indifferent to the suffering caused by such an act of destruction.

8. Darwin, C., *The Origin of Species* (New York: Gramercy, 1995).
9. Crick, F., *What Mad Pursuit: A Personal View of Scientific Discovery* (New York: Basic Books, 1988), pp. 137–42.
10. Gould, S.J., *The Panda's Thumb: More Reflections in Natural History* (New York: W.W. Norton, 1980).
11. See Monod, J., *Chance and Necessity* (Wainhouse, trans., London: Fontana, 1974) for a very frank and forthright discussion of the consequences of this possibility.
12. Huxley, T.H., "Evolution and Ethics", p. 44 (reproduced in Paradis and Williams, eds., *Evolution and Ethics: T.H. Huxley's Evolution and Ethics, with New Essays on its Victorian and Sociobiological Context*, Princeton: Princeton University Press, 1989). Huxley's view was that "ethical nature" contradicted and struggled with "cosmic nature" in order to curb the cosmic process and compensate for the lack of the qualities that best fitted us for success within this process. He prescribed (p. 182) that human beings should be "perpetually on guard against the cosmic forces, whose ends are not his ends".
13. Spencer, H., *Social Statics*, p. 414 (New York: Appleton, 1864) asserted that it is best that the poor and unfit should die "to make room for better". Of course, as Kropotkin argued, it does not necessarily follow from Darwin's theory of evolution that competition between individuals is the only strategy for survival. His observation of many instances of cooperation between members of the same species in Siberia, one of the harshest environments on Earth, showed how there are natural selections in favour of social, altruistic, and even ethical behaviour. See Kropotkin, P.A., *Mutual Aid: A Factor of Evolution* (Freedom Press, 1987). Also see Marguilis, L. and Sagan, D., *Slanted Truths: Essays on Gaia, Symbiosis and Evolution* (New York: Copernicus Books, 1997) for some contemporary discussions by evolutionary biologists for whom cooperation is considered to be the dominant evolutionary mechanism. However, whichever interpretation of the theory of evolution we prefer, concepts of struggle and opposition are utilised to represent the natural state.
14. This is not to suggest that we are the only beings that have purposes (and, consequently, might experience events in terms of good or bad luck). It may well be possible that there are many animals in this world that have a sense of their own purpose and the value of other entities for those purposes. Even if it is a prerequisite of the definition of purpose that it is a property of a mind, requiring some degree of forethought, there is still no reason to believe that the possession of a mind is an exclusively human characteristic.
15. Rogers, K., *On the Metaphysics of Experimental Physics* (Basingstoke: Palgrave Macmillan, 2005).
16. Modern science is the realisation of an ancient dream. In Thucydides' *History of the Peloponnesian War* (Hobbes, trans., Chicago University Press, 1989), art (*techne*) is represented as the means by which human beings can struggle against natural catastrophes and overcome the harshness of Nature. The Babylonian *Enuma elish* is another example of a creation myth that depicts "Man" (the hero Marduck, in this case) in battle with Nature as primordial chaos, ultimately killing the Primordial Mother and allowing Nature to emerge as bountiful and beneficent, when dominated and controlled. See Ricoeur, P., *The Symbolism of Evil* (Boston: Beacon Press, 1967) for an

interesting discussion of this myth, as well as other discussions of the histor-ical developments and interpretations of the idea of evil.

17. Sophocles' *Ode to Man* in *Antigone* laments the tragedy of humanity that uses art *(techne)* to conquer Nature and order human life through agriculture, animal domestication, medicine, the use of mechanical devices, and by building cities, but ultimately humanity becomes dominated by art, which perpetually innovates and drives humans forward, as it creates and destroys, making a world that is beyond human control and inhumane. Human beings end up dependent on the artificial world.

18. See Shapin, S., *The Scientific Revolution* (University of Chicago Press, 1996); and Worster, D., *Nature's Economy: A History of Ecological Ideas* (Cambridge University Press, 1977) for interesting discussions about the Enlightenment vision of a human society living in harmony within natural order.

19. Of course, as Nietzsche frequently pointed out, this gross assumption ignores the fact that throughout human history lies and falsehoods have been instru-mental in the development of civilisation and power. Something does not have to be objectively true in order to bring psychological, social, and polit-ical advantages. Nor is the belief in a falsehood necessarily a disadvantage. After all, what harm results from the childhood belief in faeries, elves, or Santa Claus? Much delight and wonder arises from them.

20. For an example of a recent advocate of this view, see Wilson, E.O., *Consilience* (London: Little, Brown and Co., 1998). Also see Searle, J., *The Construction of Social Reality* (London: Penguin, 1995).

21. See Bhaskar, R., *A Realist Theory of Science* (Leeds Books, 1975), for a developed argument for the necessity of a scientific realist interpretation of experiment-ation. The ideas of stratification and ontological depth are fundamental to the intelligibility of experimentation, and, hence, a positivistic interpreta-tion of science is flawed. While I agree with Bhaskar's criticisms of posit-ivism, my criticisms of Bhaskar's realist argument can be found in *On the Metaphysics of Experimental Physics*, Chapter 2.

22. For example, see Garnder, E.J., *Principles of Genetics* (New York: John Wiley, 1975); and, Mayr, E., *Evolution and the Diversity of Life* (Cambridge, Mass.: Harvard University Press, 1976).

23. *Gaia: A New Look at Life on Earth* (Oxford University Press, 2000). See also Ayala, F.J., "The Concept of Biological Progress", in Ayala, F.J. and Dobzhansky, T. (eds), *Studies in the Philosophy of Biology* (New York: Macmillan, 1974, pp. 339–55); Valentine, J.W., *Evolutionary Paleoecology of the Marine Biosphere* (Englewood Cliffs, N.J.: Prentice Hall, 1973).

24. See Dupuy, J-P., *The Mechanization of the Mind: On the Origins of Cognitive Science* (DeBevoise, trans., Princeton University Press, 2000) for a fascinating discussion of the history of the intersection between biology, computer science, and cybernetics.

25. For example, take John Polkinghorne's representation of the stars as factories for the heavier elements necessary for the production and reproduction of life in *Beyond Science* (Cambridge University Press, 1996), p. 84. Polking-horne actually extends the metaphor of the artificial back on itself to use it to explain the natural, and then argued that this explanation shows that God must have designed and made the Universe. A similar argument was made by the nineteenth-century evangelical Christian and geologist Hugh

Miller in *The Testimony of the Rocks* (Edinburgh: Nimmo, 1857). He argued that Nature's ability to adapt itself to human needs and wants, rendering it more pleasing to our aesthetic sense and material condition, shows that Nature is bound up with its perfection in relation to conscious purpose. This argument was used to suggest that God designed Nature.

26. Kuhn, T.S., *The Structure of Scientific Revolutions* (Chicago University Press, 1962).
27. Heidegger, M., "The Age of the World Picture" (in *The Question Concerning Technology and Other Essays*, Lovitt (trans.), Harper Torchbooks, 1977, pp. 115–54.)
28. Dawkins, R., *The Selfish Gene* (New York: Oxford University Press, 1976).
29. Weinberg, S., *The First Three Minutes* (New York: Basic Books, 1977).
30. Sagan, C., *The Demon-Haunted World: Science as a Candle in the Dark* (New York: Random House, 1996).
31. Cupitt, D., *What is a Story?* (London: SCM, 1991), p. 99; see also Negrete, A. and Lartigue, C., "Learning From Education to Communicate Science as a Good Story" (*Endeavour*, Vol. 28, No. 3, September 2004).
32. For wide-ranging discussions of the term "technoscience", see Marcus, G.E., ed., *Technoscientific Imaginaries* (University of Chicago Press, 1986); Aronowitz, S., *et al.*, eds, *Technoscience and Cyberculture* (Oxford: Routledge, 1996); and Gordo-Lopez, A.J. and Parker, I., eds, *Cyberpsychology* (London: Macmillan Press, 1999).
33. Haraway, D.J., "Mice into Wormholes: A Comment on the Nature of No Nature" (in Downey, G.L. and Dumit, J. (eds), *Cyborgs & Citadels: Anthropological Interventions in Emerging Sciences and Technologies*, Santa Fe, NM: School of American Research Press, 1998, pp. 209–43), p. 210.
34. Haraway, D.J., *Primate Visions: Gender, Race, and Nature in the World of Modern Science* (New York: Routledge, 1989). See Noble, D., *Forces of Production* (Basic Books, 1984) for a discussion about how technological innovations reproduce structures of class inequalities; see also Braverman, H., *Labour and Monopoly Capitalism* (New York: Monthly Review, 1974). For insightful discussions of how technological innovations and science reproduce gender inequalities, see Schwartz Cowan, R., *More Work For Mother* (New York: Basic Books, 1983); Scott, J.W., *Gender and the Politics of History* (New York: Columbia University Press, 1988), and Merchant, C., *The Death of Nature: Women, Ecology, and the Scientific Revolution* (New York: Harper and Row, 1980).
35. Haraway, D.J., *Simians, Cyborgs, and Women: The Reinvention of Nature* (New York: Routledge, 1991).
36. Marx, K., *The Poverty of Philosophy* (Quelch, trans., New York: Prometheus, 1995), p. 1.
37. Ellul, J., *The Technological Society* (trans. Wilkinson, New York: Knopf, 1964).
38. The characterisation of rationality in modern society as being a bureaucratic evaluation of social life in terms of calculation and control was central to Max Weber's social critique in *The Protestant Ethic and the Spirit of Capitalism* (Parsons, trans., New York: Scribners, 1958).
39. Even though I have taken Ellul's idea of the technological society to be an insightful starting point in my own critique of modern society, I do not wish to discuss Ellul's Christian alternative to the technological society, except to

note that I do not share his commitment to this alternative and would rather focus my own contribution to this discussion on how we can proceed with a secular critique of the rationality of the modern world. Ellul articulated this alternative in *The Politics of God and the Politics of Man* (Grand Rapids, Mich.: Eerdmaus, 1972), and *The Ethics of Freedom* (Grand Rapids, Mich.: Eerdmaus, 1976). See also Hanks, J.M., *Jacques Ellul: A Comprehensive Bibliography* (Greenwich, Conn.: JAI Press, 1984); and, Lovekin, D., *Technique, Discourse, and Consciousness: An Introduction to the Philosophy of Jacque Ellul* (Bethlehem, Pa.: Lehigh University Press, 1991).

40. For an interesting discussion of this point, see Hughes, T.P., *Human-Built World: How To Think About Technology And Culture* (University of Chicago, 2004), pp. 110–52.

41. See Mueller, M., "Technology Out of Control" (in *Critical Review*, 1.4. 1987, pp. 24–40); and, Winner, L., *Autonomous Technology: Technics Out of Control as a Theme in Political Thought* (Mass.: MIT Press, 1977). See also Ihde, D., *Technology and the Lifeworld* (Bloomington and Indianapolis: Indiana University Press, 1990), pp. 128–44, for an interesting discussion of technical pluralism and the centrality of human choice in the development of technology. Bounded technical rationality is what Andrew Feenberg termed as "reflective technology" in *Questioning Technology* (New York: Routledge, 1999), pp. 207–10.

42. See Latour, B. and Woolgar, S., *Laboratory Life: The Social Construction of Scientific Facts* (Beverly Hills, Calif.: Sage Publications, 1979); Knorr-Cetina, K.D. and Cicourel A.V. eds, *Advances in Social Theory and Methodology: Toward and Integration of Micro- and Macro-Sociologies* (Boston, Mass.: Routledge and Kegan Paul, 1981); Hughes, T.P. (1983). *Networks of Power: Electrification in Western Society, 1880–1930* (Baltimore: Johns Hopkins University Press, 1983); Callon, M., Law, J. and Rip, A. eds, *Mapping the Dynamics of Science and Technology: Sociology of Science in the Real World* (London: Macmillan, 1986); Bijker, W.E., Hughes, T.P. and Pinch, T.J., eds, *The Social Construction of Technological Systems: New Directions in the Sociology and History of Technology* (Cambridge, Mass.: MIT Press, 1987); Latour, B., *Science in Action: How to Follow Scientists and Engineers Through Society* (Milton Keynes: Open University Press, 1987); Law, J. ed., *A Sociology of Monsters: Essays on Power, Technology and Domination* (Routledge, London, 1991); and, Bijker, W.E. and Law, J. (eds), *Shaping Technology – Building Society: Studies in Sociotechnical Change* (Cambridge, Mass.: MIT Press, 1992).

43. "Question Concerning Technology", in *The Question Concerning Technology and Other Essays* (Lovitt, trans., Harper Torchbooks, 1977).

44. "Letter On Humanism" (in *Basic Writings*, Farrel Krell, ed., London: Routledge, pp. 217–65), p. 255.

45. R.K. Merton made this point in his introduction to Ellul's *The Technological Society*, pp. viii–xiii. He pointed out that the technological society responds to the needs generated by technique, as if we were responding to immutable laws of development and "technology becomes more like the new god". As the historian and philosopher E.E. Fournier D'Albe succinctly and poetically wrote in 1926, in *Hephaestus or The Soul of the Machine* (London: Keegan-Paul), this god of the modern age is personified by Hephaestus, the ancient Greek god of fire and making.

46. Mumford, L., *Technics and Civilization* (Harcourt Brace & Company, 1963).
47. *Ibid.*, p. 6.
48. Mumford, L., *Pentagon of Power: The Myth of the Machine* (New York: Harcourt Brace Jovanovich, 1970). Dwight Eisenhower first referred to the military–industrial complex in his farewell presidential speech (1961) and warned that a state-supported large arms industry could have damaging political and economic consequences. Seymour Melman developed and supported this argument in detail when he argued that the military–industrial complex was having a damaging and corrupting effect on the capitalist market economy. See *Pentagon Capitalism: The Political Economy of War* (San Francisco: McGraw-Hill, 1970) and *The Permanent War Economy: American Capitalism in Decline* (New York: Simon and Schuster, 1985).
49. Sombart, W., *Modern Capitalism: A Historical-Systematic Presentation of European Economic Life from the Beginnings to the Present* (Brookwood, 1932); and *Weltanschuung, Science, and Economy* (Veritas Press, 1939).
50. Spengler, O., *Decline of the West* (Werner, trans., Oxford University Press, 1991), first published 1918–22, and *Man and Technics: A Contribution to a Philosophy of Life* (California: University Press of the Pacific, 2002), first published in 1931.
51. Mitcham, C., *Thinking Through Technology: The Path Between Engineering and Philosophy* (University of Chicago Press, 1994).
52. Paul Tillich pre-empted and inspired many of Heidegger and Ellul's ideas, but his influence on the humanist tradition has been overlooked. For example, see his essay "The Logos and Mythos of Technology" (first published in 1927) and "The Technical City as Symbol" (first published in 1928) in Thomas, J.M. (ed.), *The Spiritual Situation in Our Technical Society* (Macon, Georgia: Mercer University Press, 1988).
53. Borgmann, A., *Technology and the Character of Contemporary Life* (Chicago: University of Chicago Press, 1984).
54. Habermas, J., *Toward a Rational Society: Student Protest, Science, and Politics* (Shapiro, trans., Boston: Beacon Press, 1970); *Legitimation Crisis* (McCarthy, trans., Boston: Beacon Press, 1975).

2 The metaphysics of modern science

1. For example, see Greenberg, D., *The Politics of Pure Science* (New York: New American Library, 1967); Bloor, D., *Knowledge and Social Imagery* (Oxford: Routledge, 1976); Latour, B. and Woolgar, S., *Laboratory Life* (Beverly Hills, Calif.: Sage Publications, 1979); Foucault, M., *Knowledge/Power: Selected Interviews and Other Writings, 1972–1977* (Gordon, trans., ed., New York: Pantheon Books, 1980); Knorr-Cetina, K., *The Manufacture of Knowledge* (Pergamon Press, 1981); Easlea, B., *Fathering the Unthinkable: Masculinity, Scientists, and the Nuclear Arms Race* (London: Pluto Press, 1983); Collins, H.M., *Changing Order: Replication and Induction in Scientific Practice* (London: Sage Publications, 1985); Harding, J., ed., *Perspectives on Gender and Science* (London: Falmer Press, 1986); Ince, M., *The Politics of British Science* (Brighton: Wheatsheaf, 1986); Billig, D., *Arguing and Thinking* (Cambridge University Press, 1987); Galison, P., *How Experiments End* (University of

Chicago Press, 1987); Latour, B., *Science in Action* (Milton Keynes: Open University Press, 1987); Suchman, L., *Plans and Situated Actions* (Cambridge University Press, 1987); Gooding, D., Pinch, T., and Schaffer, S., eds, *The Uses of Experiment: Studies in the Natural Sciences* (Cambridge University Press, 1989); Danziger, K., *Constructing the Subject* (Cambridge University Press, 1990); Gooding, D., *Experiment and the Making of Meaning* (Dordrecht: Kluwer Academic Publishers, 1990); Pickering, A., ed., *Science as Practice and Culture* (Chicago University Press, 1992); Galison, P. and Stump, D., eds, *The Disunity of Science: Boundaries, Contexts and Power* (Stanford University Press, 1996); and, Galison, P., *Image and Logic: A Material Culture of Microphysics* (Chicago University Press, 1997).

2. For example, see White, L., *Medieval Technology and Social Change* (Oxford University Press, 1962); Merton, P.K., *Science, Technology, and Society in Seventeenth Century England* (Cambridge University Press, 1970); Mathias, P., ed., *Science and Society 1600–1900* (Cambridge University Press, 1972); Bennett, J.A., "The Mechanics' Philosophy and the Mechanical Philosophy" (in *History of Science*, 1986, pp. 1–28); Yearly, S., *Science, Technology, and Social Change* (London: Unwin Hyman, 1988); Kaufman, T.D., "Astronomy, Technology, Humanism, and Art at the Entry of Rudolf II into Vienna, 1577" (in *The Mastery of Nature: Aspects of Art, Science, and Humanism in the Renaissance*, Princeton University Press, 1993, pp. 136–50); and Long, P.O., "Power, Patronage, and the Authorship of *Ars*: From Mechanical Know-how to Mechanical Knowledge in the Last Scribal Age" (in *ISIS*, 88, 1997, pp. 1–41).

3. *MEP*, Chapter 3.

4. Aristotle, NE 6.4; Metaphysics 1.1; Rhetoric 1.2 (*The Complete Works*, 2 vols, Barnes, ed., trans., Princeton University Press, 1984). For both Plato and Aristotle, *techne* referred to the general, abstract, and communicable first principles of making and inscription in the activities of craftsmanship and art. It is how *techne* was related to *episteme* (commonly translated as "science" or "knowledge of eternal and necessary principles") that differed between Plato and Aristotle. Both Plato and Aristotle considered mathematics and mechanics as *technai* along with agriculture, building, medicine, pottery, painting, sculpture, and rhetoric. However, in Plato's works, *techne* and *episteme* were used interchangeably to characterise geometrical reasoning in particular. See *Philebus* (55c–56d), *Gorgias* (450b–c), and *Ion* (532) (*The Complete Works*, Cooper, ed., Cambridge, Mass.: Hackett Publishing, 1997), for example. The mathematical activities of numbering, measuring, or weighing were taken to be the most truly *technai* because they were taken to involve the greatest precision and were more closely associated with the activities of making that operate upon the material world. These reasoned activities operated by guiding acts of making through the use of mathematics, and the *techne* of such activities, provided a formal knowledge and rules by which material practices were performed, governed, and understood. However, in *Philebus* (56d), *epistemoi* such as arithmetic and geometry were distinguished from *technai* such as carpentry because the former deals with abstract numbers and proportions whereas the latter uses numbers and proportions to deal with materials. In *The Statesman* (258e), *episteme* was used to denote pure theory or any knowledge that did not

relate to the material world in a practical manner. *Episteme* was reserved for knowledge learnt for its own sake, whereas *techne* was always directed towards the production of something else.

5. *NE* 6.4.1140a11.
6. See *Metaphysics* 7.8.1033b20–1034a7 and 7.9.1034a10–11. For Aristotle, no two lumps of clay were alike and a potter cannot make the same pot twice. *Hyle* was the particularity of any particular lump of clay and did not refer to the clay-like properties of the substance called "clay". It referred to the way that each and every pot, as well as the experience of making them, is different; even though all pots are all made out of the same substance in accordance with the same *techne*.
7. *Physics* 2.2.194a23.
8. *NE* 2.9.1109b23.
9. *NE* 2.1.1103a35 and *Metaphysics* 1.1.980b25ff.
10. *Metaphysics* 7.9.1034a10–11.
11. *Physics* Bk.2 and *Metaphysics* Book IV. He used *techne* to elucidate his conception of *phusis* as teleological (requiring *tuche*, meaning luck or chance, as a third explanatory concept).
12. See Waterlow, S., *Nature, Change, and Agency in Aristotle's Physics* (Oxford: Clarendon Press, 1982); and, Daston, L. and Park, K., *Wonders and the Order of Nature* (New York: Zone Books, 1998) for further discussion of this point.
13. *Politics* 11252b1–5. Plato was critical of the relation between mimicry (*mimesis*) and art (*Rep.* Bk. X, 596b–598c) due to the fact that the artist is a creator of an illusion (601c9) who leads the viewer away from reality. Plato described the artist as a stage magician who uses tricks (*mechanai*) in order for us to accept his imitation of Nature as being the real thing. Art is mistrusted. Art is a counterfeit.
14. The medieval Aristotelian categories of the academic disciplines included all productive pursuits, including mechanics, under the general rubric of "arts". See Weisheipl, J.A., "The Nature, Scope, and Classification of the Sciences" (in Lindberg, D.C., ed., *Science in the Middle Ages*, Chicago University Press, 1978, pp. 461–82) for further discussion and references.
15. See Pérez-Ramos, A., *Francis Bacon's Idea of Science and the Maker's Knowledge Tradition* (Oxford: Clarendon Press, 1988); and, Rossi, P., *Philosophy, Technology, and the Arts in the Early Modern Era* (Attonasio, trans., New York: Harper and Row, 1970), pp. 137–45.
16. *The New Organon* (Jardine and Silverthorne, eds, Cambridge University Press, 2000), pp. 20–1.
17. *Ibid.*, pp. 69–70.
18. *Ibid.*, p. 100.
19. *MEP*, Chapter 3.
20. See also Smith, P.H., *The Body of the Artisan: Art and Experience in the Scientific Revolution* (University of Chicago Press, 2004) for an interesting discussion of this point.
21. *New Organon*, p. 33. Note that the word "results" is a translation of the Latin *opera*, also meaning "effects" or "work". It is a derivative from *operatio*, translated as "operation" or "practice".
22. Mechanics was established as a mathematical science in sixteenth-century Italy through the influence of the University at Padua. Since the fourteenth

century, Padua had been a centre for mathematical subjects (including astronomy, astrology, geometry, optics, and geography) and was the first Italian university in the sixteenth century to offer lectures in mechanics from the chair of mathematics. Mechanics was first introduced at Padua in the 1560s in the form of lectures on Aristotelian mechanics. See *MEP*, Chapter 3 for discussion and references.

23. Moletti's arguments can be found in *In librium mechanicorum Aristotelis expositio tumultaria et ex tempore*. Milan, Biblioteca Ambrosiana MS. S 100.

24. See *MEP*, Chapter 3. See also McMullin, E., ed., *Galileo: Man of Science* (New York: Basic Books, 1967), pp. 256–92.

25. *On Motion and Mechanics* (trans., Drabkin and Drake, Madison: University of Wisconsin Press, 1960), p. 421.

26. This is evident from Galileo's use of a pendulum to demonstrate his theory of motion (*On Motion and Mechanics*, pp. 152–3); an astronomical sphere to demonstrate his theory about the Sun's rotation (pp. 348–9); and steelyards and balances to demonstrate his theory of free-fall (pp. 213–4). Having chosen the balance as fundamental and used it to derive the laws for an inclined plane, the lever, the windlass, the capstan, the pulley, and the screw, Galileo constructed a "dynamic equilibrium" method as the basis of his physics. He used this method in his treatment of hydrostatic phenomena in *Discourse on Floating Bodies* (1612) and in his *Dialogue on the Two Chief World Systems* (1632). He used this method to describe motion as separated into two independent axes, horizontal and vertical motion, to describe the fall of a body from a moving point as that of a parabola. He then rhetorically used this to argue that the Earth could revolve around the Sun (without the breath being snatched from our mouths, nor birds being flung from out of the sky). In *Dialogues Concerning Two New Sciences* (1638), he concentrated on explaining natural motion using the inclined plane. This involved using the pendulum and the balance as exemplars for his description of all natural motion and his thought experiments.

27. Bridgman, P.W., *The Logic of Modern Physics* (New York: Macmillan, 1928).

28. This is most evident in *Dialogues Concerning Two New Sciences* (Crew and de Salvio, trans., Evanston: Northwestern University Press, 1914) within which Galileo proposed the derivation of "most other mechanical devices" in terms of "the Law of the Lever" (pp. 110–2), defined his notion of force (*forza*) in terms of the lever as "mechanical advantage" (p. 124), and proposed the balance as the basic explanatory trope. Any external force could itself be simply described by mathematically projecting the lever as a half-balance.

29. See Machammer, P., "Galileo's Machines, His Mathematics, and His Experiments" (in *The Cambridge Companion to Galileo*, Machammer, ed., Cambridge University Press, 1998), for further discussion of how the balance, as a metaphor and a model, became central to every physical explanation and law.

30. See Osler, M., *Divine Will and the Mechanical Philosophy: Gassendi and Descartes on Contingency and Necessity in the Created World* (Cambridge University Press, 1994), Chapter 5; Shea, W.R., *The Magic of Numbers and Motion: René Descartes' Scientific Career* (Mass.: Science History Publications, 1991); and Heidegger, M., "Modern Science, Metaphysics, and Mathematics" (in *Basic Writings*, Farrel Krell, ed., London: Routledge, 1999),

pp. 267–306, and "On the Essence and Concept of *Phusis* in Aristotle's *Physics* Book I" (in McNeill, ed., *Martin Heidegger: Pathmarks*, Sheenan, trans., Cambridge University Press, 1998), pp. 183–230, for further discussions of this point.

31. Descartes, R., *Principles of Philosophy* (Miller and Miller (eds, trans.), Dordrecht: Reidel, 1983), p. 285. See the revised Adam and Tannery edition of *Oeuvres de Descartes* (Paris: Vrin/C.R.N.S., 1966–76, II, pp. 541–44) for a letter that Descartes wrote to Florimond de Beaune (dated April 1639) in which he described the new physics as "merely mechanics". See *MEP*, Chapter 3, for discussion.

32. See *Principles*, pp. xxvi–xxvii and p. 85. See Garber, D., "Science and Certainty in Descartes" (in *Descartes: Critical and Interpretive Essays*, Hooker (ed.), Baltimore: John Hopkins University Press, 1978, pp. 114–151) and Clarke, D.M., *Descartes' Philosophy of Science* (Pennsylvania State University Press, 1982) for further discussion of this point.

33. For a detailed discussion of Descartes' concept of living beings as mechanistic automata, see Des Chene, D., *Spirits and Clocks: Machine and Organism in Descartes* (Ithaca: Cornel University Press, 2001).

34. *Discourse on Method and The Meditations* (Sutcliffe, trans., Penguin Books, 1968), p. 78.

35. *Ibid.*, p. 91.

36. See Schaffer, S., "Glass works: Newton's prisms and the uses of experiment" (in Gooding *et al.* eds, *The Uses of Experiment*, Cambridge University Press, 1989), pp. 67–104, for a detailed discussion and references.

37. See *MEP*, Chapter 3, for discussion and references.

38. See Newton's "Preface to the 1st Edition" of *Principia*, pp. xvii–xviii and pp. 398–9 (*The Mathematical Principles of Natural Philosophy and The System of the World*, 2 volumes, Cajori, ed., trans., University of California Press, 1962).

39. "The Origin of Forms and Qualities According to the Corpuscular Philosophy" (1672) (vol. 3 of *The Works of the Honourable Robert Boyle*, Birch, ed., 6 volumes, Hildesheim: Georg Olms, 1965), p. 13; Shapin, S. and Shaffer, S., *Leviathan and the Air-Pump: Hobbes, Boyle, and the Experimental Life* (Princeton University Press, 1985); and *MEP* Chapters 3 and 4.

40. Preface, *Micrographia* (London, 1665). See R.S. Westfall, "Robert Hooke, Mechanical Technology, and Scientific Investigation" (in Burke, ed., *The Uses of Science in the Age of Newton*, Berkeley, Calif.: Berkeley University Press, 1983), pp. 85–110; and *MEP*, Chapter 4.

41. Carlino, A., *Books of the Body: Anatomical Ritual and Renaissance Learning* (Anne and John Tedeschi, trans., University of Chicago Press, 1999); Persaud, T.V.N, *A History of Anatomy: The Post Vesalian Era* (Spring Field, Il.: Charles Thomas, 1997); Sawday, J., *The Body Emblazoned: Dissection and the Human Body in Renaissance Culture* (Oxford: Routledge, 1996); and Whitteridge, G., *William Harvey and the Circulation of the Blood* (New York: American Elsevier, 1971).

42. For detailed discussions, see Allen, G.E., *Life Science in the Twentieth Century* (New York: Wiley, 1975); Sayre, A., *Rosalind Franklin and DNA* (New York: Norton, 1975); Gould, S.J., *Ontogeny and Phylogeny* (Cambridge, Mass.: Harvard University Press, 1977); Barnes, B. and Shapin, S., *Natural Order: Historical Studies of Scientific Culture* (Beverly Hills, Calif.: Sage, 1979);

Mayr, E. and Provine, W., eds, *The Evolutionary Synthesis* (Cambridge, Mass.: Harvard University Press, 1980); Farber, P., "The Transformation of Natural History in the Nineteenth Century", *Journal of the History of Biology*, 1982, 15, pp. 145–52; Sapp. J., "The Struggle for Authority in the Field of Heredity, 1900–1932: New Perspectives on the Rise of Genetics", *Journal of the History of Biology*, 1983, 16, pp. 311–342; Kevles, D.J., *In the Name of Eugenics: Genetics and the Uses of Human Heredity* (New York: Knopf, 1985); Pauly, P.J., *Controlling Life: Jacques Loeb and the Engineering Ideal in Biology* (New York: Oxford University Press, 1987); Haraway, D., *Primate Visions: Gender, Race, and Nature in the World of Modern Science* (New York: Routledge, 1989); Foucault, M., *The Order of Things: An Archaeology of the Human Sciences* (New York: Vintage Books, 1994); Lee, K., *Philosophy and Revolutions in Genetics* (Palgrave Macmillan, 2002); and, Trusted, J., *Beliefs and Biology* (Basingstoke: Palgrave Macmillan, 2003).

43. For an interesting history of the emergence of ecological science as an empirical science, see Worster, D., *Nature's Economy: The Roots of Ecology* (Garden City, New York: Anchor Books, 1979).

44. Dupuy, *The Mechanization of the Mind: On the Origins of Cognitive Science* (DeBevoise, trans., Princeton University Press, 2000).

45. See *MEP*, Chapters 4 and 5, for a detailed discussion of this.

46. Hacking, I., *Representing and Intervening: Introductory Topics in the Philosophy of Natural Science* (Cambridge University Press, 1983).

47. Gooding, D., *Experiment and the Making of Meaning* (Dordrecht: Kluwer Academic Publishers, 1990).

48. Cartwright, N., *The Dappled World: A Study of the Boundaries of Science* (Cambridge University Press, 1999).

49. Bhaskar, R., *A Realist Theory of Science* (Leeds Books, 1975).

50. *Ibid.*, pp. 87–90.

51. *Ibid.*, p. 112.

52. Kroes, P. and Meijers, A., eds, *The Empirical Turn in the Philosophy of Technology: Research in Philosophy and Technology, Volume 20* (Oxford: Elsevier Science, 2000). See also Rothbart, D., "The Dual Nature of Chemical Substance", *Techne*, 6:2, 2002, pp. 28–33.

3 The technological society

1. Friedrich Dessauer (1881–1963) was a proponent of the view that the technological imperative to transform the world for the better was a moral imperative and brought human beings into contact with things-in-themselves. The creative process of invention creates existence out of essence and the result is a working, practical solution to a problem. The autonomous, world-transforming consequences of modern technology demonstrate its transcendent moral value and, for Dessauer, technology had become a new way for human beings to exist in the world that fitted into the Kantian understanding of a categorical imperative. Dessauer was a devout Catholic and a German philosopher, who wrote several books on theology, defended technology in the strongest possible terms, sought to open up dialogue with existentialists, social theorists, and theologians, and opposed

Hitler. For this last act he was forced to flee Germany. For discussion of Dessauer, see Mitcham, *Thinking Through Technology*, pp. 29–33; and Klaus Tuchel, "Friedrich Dessauer as Philosopher of Technology: Notes on His Dialogue with Jaspers and Heidegger" (in *Research in Philosophy and Technology*, vol. 5, pp. 269–80, Durbin (ed.), Greenwich, Conn.: JAI Press, 1982).

2. *Technological Society*, pp. 38–45.
3. *Ibid.*, p. 86.
4. *Ibid.*, p. 5 and 21. For Ellul's discussions about the relationships between Nature, magic, and technique in primitive societies, see pp. 24–7, 36–7, and 64–9. See *MEP*, Chapter 1, for a discussion of the parallels between Ellul and Heidegger, as well as my criticisms of Ellul's analysis of the distinction between science and technology.
5. *Ibid.*, p. 24.
6. *Ibid.*, pp. 5 and 53. Ellul cites Mumford's *Technics and Civilization* as support for this claim, but as I shall explain in the next section, Mumford's view on the relationship between technology and capitalism was more complicated and critical than this.
7. *Ibid.*, pp. 104–5.
8. *Ibid.*, p. 184.
9. *Ibid.*, p. 197.
10. *Ibid.*, p. 144.
11. *Ibid.*, pp. 200–201.
12. *Ibid.*, p. 198 and pp. 236–7.
13. *Ibid.*, p. 81.
14. Ellul, J., *The Technological Bluff* (Bomiley, trans., Grand Rapids, Mich.: Eerdmans, 1990).
15. See Landes, D., *Bankers & Pashas: International Finance and Economic Imperialism in Egypt* (Cambridge, Mass.: Harvard University Press, 1959) for an interesting account of the emergence of capitalism.
16. *Technics and Civilisation*.
17. See Lovins, A.B., *Soft Energy Paths: Toward a Durable Peace* (New York: Harper & Row, 1979), for an interesting discussion about how alternative technologies were suppressed by the oil and car industries. Over 26 years later, we see that not much has changed.
18. *T&C*, pp. 167–8. Benjamin Franklin proposed a method to utilise unburnt carbon in coal smoke by recycling and burning it a second time in the furnace. This method was never used. Steam power is highly inefficient. About 90 per cent of the heat produced escapes in the steam and smoke. The steam engine was also extremely noisy. James Watt's efforts to improve the efficiency of the steam engine and reduce its noise were dismissed as being too expensive to implement.
19. *Ibid.*, p. 181.
20. *Ibid.*, p. 185.
21. *Ibid.*, p. 285.
22. *Ibid.*, p. 255.
23. See Stiegler, B., *Technics and Time, 1: The Fault of Epimetheus* (Beardsworth and Collins, trans., Stanford University Press, 1998) for an interesting discussion of how this made the temporality of life meaningless by construing all labour and technical action in terms of its value, measured in terms of time and function, for the future.

24. For Marx, the social mode of production is the dominant historical factor and, hence, the hand-mill was bound up with selfdom in a feudal system, and the steam mill with the wage-labour of industrial capitalism. See *The Poverty of Philosophy*, Chapter II. For discussions of the technological determinism inherent to Marxist ideology, see D. Mackenzie, "Marx and the Machine", *Technology and Culture*, 25, July 1984, pp. 473–502; and Winner, *Autonomous Technology*, pp. 65–75.
25. *Capital* I (Engels, ed., More and Aveling, trans., New York: International Publishers, 1967), p. 441.
26. *T&C* p. 263.
27. *Ibid.*, p. 258.
28. *Ibid.*, pp. 264–6.
29. *Ibid.*, p. 213.
30. *Ibid.*, p. 211.
31. In the *Dialectics of Nature*, Engels extended Marx's theory beyond its limits as a method for understanding history and social reality, based upon scientific realism, and developed the dialectical method from the results of contemporary natural science. Leon Trotsky stated that human beings should use science and technology to rearrange mountains and rivers, to improve on Nature and transform the world according to human choice. See *Literature and Revolution* (University of Michigan, 1960), pp. 251–3.
32. As Langdon Winner pointed out in *Democracy in a Technological Society* (Dordrecht: Kluwer Academic Publishers, 1992), p. 3, "The fact that various communist and socialist alternatives have now fallen into disfavour has not eliminated the fundamental problems that gave rise to them."
33. *Capital* I, p. 578.
34. *One-Dimensional Man* (Boston: Beacon Press, 1991) was first published in 1964. It presupposed the ideals and vision of a free and rational society articulated in *Reason and Revolution* (first published in 1941) and *Eros and Civilization* (first published in 1955). In response to his critics, Marcuse modified and developed his description and analysis of one-dimensional society in his later works *An Essay on Liberation* (1969) and *Counterrevolution and Revolt* (1972), but his basic position against social forms of domination and conformity remained the same until his death in 1979.
35. *One-Dimensional Man*, pp. 226–7.
36. *Ibid.*, p. 17.
37. Marcuse was critical of the "one-dimensional" trajectory of both industrial capitalist and communist countries. In his book *Soviet Marxism* (New York: Columbia University, 1958), Marcuse was highly critical of the USSR, but he rejected the Western countries' Cold War propaganda that claimed the moral superiority of capitalism over communism.
38. Marcuse considered the dialectical method as demonstrated in the dialogues of Plato to have been one of the most influential sources for Renaissance metaphysics.
39. *ODM*, p. 153.
40. *Ibid.*, pp. 156–7.
41. *Ibid.*, p. 158.
42. *Ibid.*
43. Marcuse accepted Husserl's interpretation of Galileo's physics having emerged from and referred back to a pre-scientific world of practice and

practical arts. In my view, this is correct but fails to address the question of how it was possible for Galileo to epistemologically connect the practical arts with natural philosophy. Husserl failed to show how Galileo was able to connect geometry with practical activity in such a way as to correlate ideational truth with empirical reality through mechanical devices.

44. In the sense of being defined in terms of (1) the validation of cognitive thought by the facts of experience; (2) the orientation of philosophical thought to the physical sciences as the model of scientific thought; (3) the belief that progress depends upon this orientation; (4) the rejection of all metaphysics; and (5) the object world is understood in terms of its instrumentality. See *ODM*, p. 172.

45. *Ibid.*, p. 136, f.n. 4.

46. Marcuse, H., *An Essay on Liberation* (Boston: Beacon Press, 1972), p. 19.

47. *T&C* p. 269.

48. See Horkheimer, M. and Adorno, T., "The Culture Industry: Enlightenment as Mass Deception", in *Dialectic of Enlightenment* (Calif.: Stanford University Press, 2002), pp. 94–136; and also Ellul, J., *Propaganda: The Formation of Men's Attitudes* (New York: Vintage Books, 1964), for fascinating analyses and discussions of conformity and mass democracy.

4   The confrontation with nature

1. In *Leviathan* (London: Penguin Classics, 1982), Hobbes asserted that we could only have knowledge of those things that we made ourselves, and, therefore, we could only have certain knowledge about society and its material products. In the materialism of Hobbes, the origin of all knowledge, including mathematical knowledge, is in the world accessible to the senses. On such an account, human desires and actions are the consequences of matter and motion governed by natural laws. Freedom and power are identical. The representations of the objects of the world that we induce or infer on the basis of our senses are tested on the basis of their practical utility in achieving the success of our intentions. The instrumentality of our representations for achieving our purposes is taken to be proof of their connection with reality. Such representations are falsified if and when their use is taken to be the cause of our failure to achieve our ends. In this way, according to Hobbes, all the qualities of a thing are knowable through perception and the "thing in itself" is the sum total of all these perceptions plus the intuition that the thing must exist without us. In other words, whilst Hobbes intuitively acknowledged the independent existence of the world from perception (given that perception is nothing more than the consequences of matter and motion governed by natural laws), the properties of things are exactly those properties that can be represented and are practically useful to human beings. Thus human history becomes identical with natural history in so far as both are descriptions of the motion of matter in accordance with immutable natural laws. For further discussion, see Mintz, S.I., *The Hunting of the Leviathan: Seventeenth Century Reactions to the Materialism and Moral Philosophy of Thomas Hobbes* (Cambridge University Press, 1969).

2. Horkheimer and Adorno, *Dialectic of Enlightenment*, p. 17.
3. *Ibid.*, p. xiv. See also Horkheimer, M., *Critique of Instrumental Reason: Lectures and Essays since the End of World War II* (O'Connell *et al.*, trans., New York: Continuum, 1974); and Adorno, T., *The Stars Down to Earth, and Other Essays on the Irrational in Culture* (Crook, ed., London: Routledge, 1994).
4. *Techne* was the possession of the most helpless, unshod, unarmed, unclad, but highest animal who could, through *techne*, turn this weakness around, take advantage of *phusis*, and even complete that which *phusis* left incomplete (*Physics* 2.8; *Politics* 1337a1–2). For Aristotle, *techne* was rooted in and a completion of *phusis* to the extent that even human nature was completed by *techne* through medicine, crafts, and politics (*Physics* 193b10 and 2.1.193a12–17; *Politics* 1.2.1253a2).
5. *On the Natural Faculties* (Brock, A.J., trans., London: Heinemann, 1947).
6. *Natural History* (Healy, trans., London: Penguin Classics, 1991), Bk. 21, XXII.
7. See *MEP*, Chapter 3, for discussion and references. See also Laird, W.R., "The Scope of Renaissance Mechanics", *Osiris*, 2nd series, 2 (1986), pp. 43–68; and Hackett, J., ed., *Roger Bacon and the Sciences* (Leiden: Brill, 1997) for further discussion.
8. Newman, W.R., *Promethean Ambitions* (University of Chicago Press, 2004).
9. *Ibid.*, p. 33. Of course, some readers will point out that alchemy could not fulfil its promises and it was a pseudo-science, whereas genetics is a proper science. While I acknowledge this point, it seems to me, on one side, that it is somewhat Whiggish because the early alchemists did not know what the limits of their art were and which of their claims were fantastic, thus the purpose of their efforts were to discover these things by trying to realise their claims, and, on the other side, is also somewhat presumptive because it assumes that genetic engineering will be able to realise the equally fantastic claims made about its potential. It may well be the case that genetic engineering will not fulfil its promises and, hence, could more similar to alchemy than some people would like to think.
10. The Jewish golem is distinct from the Arabic homunculus created by alchemy. The golem is created from the magical/divine power of words, whereas the homunculus is created from an alchemical process involving sperm being mixed with other substances and being inserted into an artificial matrix. The idea of the golem was to provide a parable of human folly and limitation (in some sense the golem is always a failure), whereas the homunculus was used to describe an ideal conception of human being in terms of intellect and moral character (as well as often being capable of magical powers). The golem is created by the magic of words, but lacks speech and intelligence, whereas the homunculus lacks the imperfections of natural being and thereby struggles to live in the imperfect natural world. For further discussion, see Pines, S., "The Origin of the Tale of Salaman and Absal: A Possible Indian Influence" (in Pines, *Studies in the History of Arabic Philosophy*, Jerusalem: Magnes Press, 1996, pp. 343–53); Idel, M., *Jewish Magical and Mystical Traditions on the Artificial Anthropoid* (Albany: State University of New York Press, 1990); and Newman, *Promethean Ambitions*, Chapter 4.
11. *PA*, pp. 301–2.
12. For contemporary examples of this utopian vision of the biotechnological society, see Naam, R., *More Than Human: Embracing the Promise of Biological*

Enhancement (New York: Broadway, 2005); and Easterbrook, G., *A Moment on the Earth: The Coming Age of Environmental Optimism* (London: Penguin, 1996). Current developments in biotechnology have recently led lesbian proponents of gynogenesis to advocate the use of biotechnology to artificially produce a completely female humanity on the premise that such a society would be a considerable advance over the natural state of affairs. For an example, see Sourbut, E., "Gynogenesis: A Lesbian Appropriation of Reproductive Technologies" (in Lykke, N. and Braidotti, R., eds, *Between Monkeys, Goddesses, and Cyborgs: Feminist Confrontations with Science, Medicine, and Cyberspace*, London: Zed Books, 1996, pp. 227–41).

13. *PA*, Chapter 2.
14. *Meteorology* (IV 3 381b3–9). See *PA*, pp. 64–6, for discussion.
15. *PA*, pp. 145–63, for discussion. See also Dalton, L. and Park, K., *Wonders and the Order of Nature* (New York: Zone Books, 1998).
16. Both Boyle and Newton considered alchemy to be a legitimate way of investigating natural processes and laws. For further discussion of the alchemical interest and experiments of Boyle and Newton, see Principe, L., *The Aspiring Adept: Robert Boyle and his Alchemical Quest* (Princeton University Press, 2000) and Dobbs, B.J.T., *The Janus Face of Genius: The Role of Alchemy in Newton's Thought* (Cambridge University Press, 2002). John Locke wrote down a method to spontaneously generate toads or serpents from a rotting goose or duck carcase (MS Locke C44, Bodleian Library, Oxford University). This method was taken from Robert Boyle's notes on palingenesis. See Boyle's essays "Essay on the Holy Scriptures" and "Some Physio-Theological Considerations About the Possibility of the Resurrection" in *The Works of Robert Boyle* (Hunter and Davis, eds, London: Pickering & Chatto, 2000). For interesting discussions of the historical connections between science and magic, see also Thorndike, L., *History of Magic and Experimental Science* (New York: Columbia University Press, 1934); Yates, F., *Giordano Bruno and the Hermetic Tradition* (Chicago University Press, 1964); Rossi, B., *Francis Bacon: From Magic to Science* (Rabinovich, S., trans., London: Routledge & Keegan Paul, 1968); and Copenhauer, B., "Natural Magic, Hermetism, and Occultism in Early Modern Science" (in Lindberg, D.C. and Westman, R.S., *Reappraisals of the Scientific Revolution*, Cambridge University Press, 1990, pp. 261–301).
17. *PA*, pp. 256–71.
18. Moran, B., *Distilling Knowledge: Alchemy, Chemistry, and the Scientific Revolution* (Harvard University Press, 2005).
19. Newman did acknowledge that mechanics was an important influence on Robert Boyle, in reference to a paper by Margaret Cook ("Divine Artifice and Natural Mechanism: Robert Boyle's Mechanical Philosophy of Nature" in *Osiris*, 2nd series, 16, 2001, pp. 133–50), but he neglected to relate that influence to practical activity and the growing dominance of maker's knowledge as a knowledge of natural laws.
20. *The New Atlantis and The Great Insauration* (Wheeling, Il.: Harlan Davidson, 1989). For further discussion, see Price, B., ed., *Francis Bacon's The New Atlantis: New Interdisciplinary Essays* (Manchester University Press, 2003).
21. According to Augustine's conception of "original sin" and "free will", the cruelty and suffering inflicted upon human beings by Nature was taken

to be a consequence of Adam's disobedience and fall from grace, but on such an interpretation there was not any notion of progress on Earth. The sublime Parousia of the grace of Eden can only be regained through redemption and it cannot be achieved through human labour and struggle. See Hefner, P., *The Human Factor: Evolution, Culture, and Religion* (Minneapolis: Fortress Press, 1993); and Brooke, J.H. and Cantor, G., *Reconstructing Nature: The Engagement of Science and Religion* (New York: Oxford University Press, 1998), for contemporary Christian theological discussions of this problem. See Orange, D., "Oxygen and the One God" (*History Today*, 24, 1974, pp. 773–81) for a discussion of how Joseph Priestly subsumed his notion of Providence under concepts of scientific and technological progress towards a better world. Also see Horkheimer and Adorno, *Dialectic of Enlightenment*, for an interesting discussion of the use of science to achieve a god-like mastery and gaze over Nature.

22. In his essay "Nature", first published in 1874, Mill argued that everything that is artificial is also natural.

23. "The Question Concerning Technology" (in *The Question Concerning Technology and Other Essays*, Lovitt, trans., Harper Torchbooks, 1977, pp. 3–35).

24. As well as QCT, see "Modern Science, Metaphysics, and Mathematics" (in *Basic Writings*, Farrel Krell, ed., London: Routledge, pp. 267–306). See also, *MEP*, Chapters 3 and 4, for critical discussion of Heidegger's philosophy of physics.

25. See Aristotle (N.E. Bk. 6) for his distinctions between the intellectual virtues of *episteme, techne, sophia, nous,* and *phronesis.* The criticism of the dominance of the intellectual virtue of *techne* in modern society was central to the critical analyses of modern society presented by Arendt, H., *The Human Condition* (Chicago University Press, 1958); Habermas, J., *Theory and Practice* (Viertel, trans., Boston: Beacon, 1973); and Dunne, J., *Back to the Rough Ground: "Phronesis" and "Techne" in Modern Philosophy and in Aristotle* (Indiana: University of Notre Dame Press, 1993).

26. QCT, pp. 6–13.

27. For further discussion of this point, see Rothenberg, D., *Hand's End: Technology and the Limits of Nature* (Los Angeles: University of California Press, 1993).

28. *MEP*, Chapters 4 and 5.

29. "Letter on Humanism" (in *Basic Writings*, Farrel Krell, ed., London: Routledge, 1999, pp. 217–65), p. 259, footnote.

30. Geyer, F. and van der Zouwen, J., eds, *Sociocybernetics: Complexity, Autopoiesis, and Observation of Social Systems* (Westport, Conn.: Greenwood Press, 2001).

31. For example, see Levidow, L. and Young, B., eds, *Science, Technology, and the Labour Process: Marxist Studies* (London: CSE, 1981).

32. Critical realism presumes mechanical realism and the rationality of the societal gamble, given that it assumes that scientific knowledge is good and that human emancipation depends on a scientific and dialectical relation between transitive human activity and intransitive natural structures and laws. This philosophical movement grew out of the writings of Roy Bhaskar and owes a considerable debt to Marxism. Of particular importance are Bhaskar's books *Scientific Realism and Human Emancipation* (London: Verso

1986); *The Possibility of Naturalism* (New York: Harvester Wheatsheaf, 1989); and *Dialectic: The Pulse of Freedom* (London: Verso, 1993).

33. Kolakowski, L., *Marxism and Beyond: On Historical Understanding and Individual Responsibility* (Peel, trans., London: Pall Mall Press, 1968), p. 66.
34. Lukács, G., *History and Class Consciousness* (Livingston, trans., London: Merlin Press, 1967).
35. *Ibid.*, p. 231.
36. *Ibid.*, pp. 157–8.
37. "What is Orthodox Marxism?" *ibid.*, pp. 1–25.
38. See *Dialectics of Nature* (Chapter IX).
39. *H&CC*, p. 83.
40. Passmore, J., *The Perfectibility of Man* (London: Duckworth, 1970), p. 240.
41. For example, see Plumwood, V., *Feminism and the Mastery of Nature* (London: Routledge, 1993) and Lee, K., *Philosophy and Revolutions in Genetics* (Basingstoke: Palgrave Macmillan, 2003).
42. For critical discussions of whether biotechnology is actually based on good science, see Lee, K., *Philosophy and Revolutions in Genetics* (Palgrave Macmillan, 2003); and Ho, Mae-Wan, *Genetic Engineering: Dream or Nightmare? The Brave New World of Bad Science and Big Business* (Bath: Gateway, 1998).
43. Marshall McLucan, in *Understanding Media* (New York: McGraw Hill, 1964), p. 46, said technology reduces us to being "the sex organs of the machine world". Perhaps the development of biotechnology will make us redundant in that capacity too.
44. Mannheim, K., *Ideology and Utopia* (Wirth and Shils, trans., New York: Harcourt Brace, 1936).
45. *T&C*, pp. 365–6.
46. See Scott, J., *Seeing Like a State: How Certain Schemes to Improve the Human Condition Have Failed* (New Haven: Yale University Press, 1998); and Josephson, P., *Industrialized Nature* (Washington, D.C.: Shearwater Books, 2002) for interesting discussions and examples.
47. *ODM*, p. 79.

5 Labour and the lifeworld

1. See Hobbes, *Leviathan*, especially Chapter 5.
2. Lukács, G., *The Ontology of Social Being: 3: Labour* (Fernbach, trans., London: The Merlin Press, 1978).
3. *Ibid.*, p. 3.
4. Of course, we can look at a stone and consider it to be beautiful, but we would be departing from a materialist interpretation of the stone if we considered its beauty to be a property of the stone.
5. See *MEP*, Chapter 5, for further discussion of how natural laws are abstracted from the labour processes involved in experimentation, and how they are used in the ongoing development and testing of physical theories.
6. Lukács discussed the origin of this concept of freedom in the philosophical work of Hegel and Engels (*OSB*, pp. 120–6). It is also central to the philosophy of Bhaskar in *Dialectic: The Pulse of Freedom*.

7. For further discussion of the relation between geometry, art, and practical activity, see *MEP*.
8. This also explains how Kuhn's conception of a scientific revolution in terms of a paradigm shift is possible. The succeeding paradigm can emerge from the practices of the former, while developing autonomy from the former, to the extent that interpretations of the two paradigms become incommensurable.
9. *OSB*, p. 52.
10. Lukács noted that it should not be surprising that the naming of objects was taken by many ancient peoples as being a magical process of obtaining mastery over objects. Lukács used the Old Testament example of Adam naming the animals as a means of obtaining mastery over them. *OSB*, p. 101.
11. Thus people commonly represent themselves as engaging in a natural carnivorous activity when they eat a hamburger or a char grilled steak, rather than representing this act as being a social mediated product of artificially transforming the digestibility and appearance of carrion (by accelerating its decay through heating and creating its aesthetic form as a meal), which has been provided by other people in accordance with complex economic activities, relations, and institutions. Of course, raw meat is also a social product requiring socially developed, differentiated skilled practices and tools to kill and butcher a living animal in order to transform into cuts of raw meat. While hunting an animal also requires socially developed and differentiated skilled practices and weapons, this is even more complex when the animal has been farmed and is often a social product of domestication and selective breeding. In a modern society, only a small proportion of people have any immediate relation with the animals that are killed for food.
12. As Lukács pointed out (*OSB*, p. 110), magic, as a means of mastering forces of Nature or obtaining favours from gods, is itself based on a model of labour. Furthermore, following Engels, Lukács suggested that the suppression of the teleology of labour is bound up with the contempt for labour expressed by the ruling class, which no longer is involved in producing and reproducing the material conditions for their existence, and, thus, equates their own purposes and organisation of society with the rational principle of a cosmic, divine, or natural order of being.
13. *Technology and the Character of Contemporary Life*.
14. Borgmann, A., *Crossing the Postmodern Divide* (University of Chicago Press, 1992).
15. *Ibid.*, p. 108.
16. Ihde, D., *Bodies in Technology* (Minneapolis: University of Minnesota, 2001).
17. Ihde was considerably influenced by Maurice Merleau-Ponty's concept of *praktognosia* as being pre-conscious, habitual, and practical knowledge that orders the motility and praxis of the body-subject without the need for reflection or mental representations (see *The Phenomenology of Perception*). This is a form of embodied knowledge that is unreflectively and unconsciously acquired through experience, habit, and non-verbal interpersonal communication (such as imitation and visualisation). This idea has considerable commonality with Michael Polanyi's concept of *tactic knowledge* (see *Personal Knowledge*).
18. *T&C*, p. 332.

19. *Ibid.*, p. 52.
20. Habermas, J., *Toward a Rational Society* (Shapiro, trans., Boston: Beacon Press, 1970).
21. Habermas, J., *The Theory of Communicative Action: Lifeworld and System: A Critique of Functionalist Reason* (McCarthy, trans., Boston: Beacon, 1984).
22. For an interesting discussion of this point, see Lie, M. and Sorensen, K., *Making Technology Our Own? Domesticating Technology Into Everyday Life* (Oslo: Scandinavian University Press, 1996).
23. *Toward a Rational Society*, p. 61.
24. Habermas, J., "Modernity – An Incomplete Project" (in Hal Foster, ed., *The Anti-Aesthetic*, Seattle, WA: By Press, 1983).
25. *The Poverty of Philosophy*, pp. 58–9.
26. *The Whale and the Reactor* (University of Chicago Press, 1986), p. 29.
27. *Autonomous Technology.*
28. Foucault, M., *Power/Knowledge: Selected Interviews and Other Writings 1972–1977* (Gordon, ed., New York: Pantheon, 1980).
29. *Ibid.*, p. 189.
30. *Ibid.*, p. 203.
31. *Ibid.*, pp. 212–13. Of course, even if we should condemn acts of revenge against members of the social elite, as being nothing more than pointless violent acts motivated by resentment and spite, it could well be argued that those people "at the top", who benefit most from society, have an ethical obligation to change the structures and mechanisms of power that reproduce social inequalities or at least refuse to participate in them. Even if these structures and mechanisms have been imposed on them from birth (as much as they have on those that are "beneath them"), it does not release them from their obligation to resist this imposition merely because they benefit from it more than others.
32. *Ibid.*, p. 98.
33. *Ibid.*, p. 214.
34. Feenberg, A., *Questioning Technology* (New York: Routledge, 1999), p. 79.
35. Feenberg frequently used Bijker's example of the bicycle to point out that the final design of the bicycle that we take for granted today was the result of a long, historical struggle between different social groups about the purpose of a bicycle, rather than a derivative product of an abstract technical rationality. While this is quite correct, in my opinion, it fails to acknowledge that a central feature of the bicycle is that it has two, circular wheels, and that there is a distinct, technical limit on the possibilities for any alternative design for a bicycle. Of course, it is possible to make a clown bicycle, with square or ovoid wheels of different size, and so on, but the reason why such a bicycle is funny is because it breaks with the technical limit of the design of an efficient and stable machine performance of a bicycle. Its humorousness is in direct proportion to the inefficiency and instability of its performance. To use another of Feenberg's favourite examples, it is clearly the case that human choices were involved in dealing with the problem of exploding boilers of the nineteenth-century American steamboats. This struggle between the politicians seeking votes, the boat owners seeking profits, the passengers seeking a means of safe travel, and crew seeking wages (and to live long enough to spend them) shows that

what constitutes "a safe boiler" design is bounded by socially defined situations and solutions. The design of a safe boiler is socially contingent, in so far that not only is every proposed design socially contingent, but the final choice is as well, as well as the need for such a design in the first place. However, minimal conditions for the design of a safe boiler are that it turns water into steam and does not explode. This does not mean that a boiler that has not exploded is a safe boiler, given that it might explode in the future, but it does mean that if it does not produce steam then it is not a working boiler, and any boiler that explodes is not a safe boiler. While the definition of the minimal conditions of the design of any machine are socially contingent, whether it satisfies those conditions is not determined only by social relations and contexts. Social construction theory ignores this asymmetry.

36. It is clear that Heidegger, Ellul, and Marcuse were all aware of this suppression of the contingency of human choices in the technological development of society. Feenberg's criticisms of Heidegger, Ellul, and Marcuse are very unfair because he accuses them of being essentialists and determinists about technology, when, in fact, it is quite obvious that they were concerned with a dominating tendency of modern technology, and all were aware that it does not have to be like this and alternatives are available to us. They simply disagreed on the alternatives.

37. *Questioning Technology*, p. 80.

38. Latour, B., "Where Are the Missing Masses? The Sociology of a Few Mundane Artifacts" (in Bijker, W. and Law, J., eds, *Shaping Technology, Building Society: Studies in Sociotechnical Change*, Cambridge, Mass.: MIT Press), p. 232.

6 Into the future

1. I do not intend any connotation of human supremacism by this remark. I simply mean that a beast does not care about his or her future, only about his or her present. It is quite possible that there are many animals on Earth (or elsewhere) that also think about their future and try to find the best course of action in the absence of knowledge. In which case, they would also not be beasts either.

2. As Borgmann argued in *Technology and Contemporary Life*, pp. 114–24, the democratic and rational revaluation of technological innovation must be based upon a community engagement with the articulation and critical assessment of the concrete proposals of technology in contrast to the current conditions of their fullness, in accordance with considerations of sufficiency for a practical appraisal of the good life. See also Borgmann, A., "Communities of Celebration: Technology and Public Life" (in Ferré, ed., *Research in Philosophy and Technology*, vol. 10, Greenwich: JAI Press, 1990, pp. 315–45).

3. Hayek, F.A., *The Road to Serfdom* (New York: Routledge, 2001).

4. There are also obvious moral criticisms about the hypocrisy of the foundation of political and economic rationality of the American form of liberal capitalism, using the prosperity of the United States of America as evidence for the success of free-market individualism, when its wealth is historically

based upon the slavery of Africans and the theft of the land of the Native Americans.

5. See Herman, E.S. and Chomsky, N., *Manufacturing Consent: The Political Economy of the Mass Media* (New York: Pantheon Books, 1988); Chomsky, N., *Media Control: The Spectacular Achievements of Propaganda* (New York: Seven Stories Press, 2002); and Ellul, *Propaganda*, pp. 90–116 and pp. 232–50, for interesting and detailed discussions. Also, as Horkheimer and Adorno argued, in *Dialectic of Enlightenment*, the modern media ("the culture industry") has degenerated into the production of social conformity and mass entertainment.

6. Thus, Mumford considered the scientific and rational development of the technological basis of society to inevitably lead to communism (*T&C*, pp. 355–6). Mumford considered his conception of communism as essentially post-Marxist because he rejected the paleotechnic values upon which he considered Marxism to be based. After he expressed admiration for "soviet courage and discipline", he rejected the idea that communism needs to adopt the methods or take the political and institutional form proscribed by Marx, Lenin, Stalin, and the Soviet Union. He stated that his notion of communism – as a universal system for distributing the essential means of life – owed more to Plato than Marx (*T&C*, p. 403). He pointed out that within many modern societies schools, libraries, universities, museums, swimming pools, hospitals, sports facilities, and public parks are supported by and available to the community at large. Emergency services, such as police, fire, and ambulance, are already provided on the basis of need rather than ability to pay and, hence, likewise could be considered as basic communism. Also, even though it takes the form of welfare, a basic communism exists in most modern countries as far as provisions for the unemployed and the elderly is concerned.

7. The centralised and dictatorial Soviet Five-Year Plans were a good example of how a reduced stock of decision-makers were a structural handicap and antithetical to the whole social process of intelligent, flexible organisation of technological innovation within a complex and unpredictable world. After Stalin imposed the centralised plan to transform a backward feudal state into a modern industrialised state, the authoritarian and ideologically driven projects implemented by Soviet engineers and architects imposed their technologies upon local communities and workers without any knowledge about the conditions and complexities of the local situation and circumstances. Such a system was unable to develop in a flexible and intelligent way because it was unable (or unwilling) to take advantage of local knowledge. Thus centralisation resulted in inefficiency, waste, famine, and environmental catastrophes. As Vitaly Gorokhov argued, using the treatment of engineers in Stalin's USSR as an example, democracy is a condition for genuine scientific and technical progress because centralised planning fails to foresee all the details in advance and, hence, generates an anarchy of *diktats* and all the inefficiencies and self-deception that follows from this. See "Politics, Progress, and Engineering: Technical Professionals in Russia" (in Winner, L., ed., *Democracy in a Technological Society*, Dordrecht: Kluwer, 1992, pp. 175–85).

8. For detailed discussion of this point, see Feenberg, A., *Critical Theory of Technology* (New York: Oxford University Press, 1991), Chapters 2 and 6; Thomas, P., *Alien Politics: Marxist State Theory Revisited* (New York: Routledge, 1994); Cunningham, F., *Democratic Theory and Socialism* (Cambridge: Cambridge University Press, 1987).
9. *Toward a Rational Society.*
10. Barber, B., *Strong Democracy: Participatory Politics for a New Age* (Berkely, Calif.: University of California Press, 1984). Ellul (in "Technology and Democracy" in Winner, ed., *Democracy in a Technological Society*, pp. 35–50) also argued that there is an absence of democracy in Western political institutions, called for the abolition of the political class of professional politicians, and advocated the replacement of the current representative system in favour of a collective and participatory form of democracy among the general citizenry.
11. On this point, I very much agree with John Rawls, *A Theory of Justice* (Cambridge, MA: Harvard University Press, 1971), p. 61.
12. Feenberg, A., *Critical Theory of Technology* (New York: Oxford University Press, 1991). Feenberg rejected the dichotomy between technological rationality and humanist values implicit in the debate between instrumentalists and substantivists. He took this distinction from Borgmann's *Technology and the Character of Contemporary Life* (p. 9) and cited Nicholas Rescher and Emmanuel Mesthene as proponents of the instrumentalist theory of technology: that technology is a neutral means to rationally satisfy ends or goals established in other social spheres. See Rescher, N., "What is Value Change? A Framework for Research" (in K. Baier and N. Rescher, eds, *Values and the Future*, New York: The Free Press, 1969), and Mesthene, E., *Technological Change* (New York: Signet, 1970). He cited Ellul, Heidegger, Habermas, and Borgmann as proponents of the substantive theory of technology: that technology is an autonomous phenomenon, overriding all traditional or non-technical values, shaping both humanity and the natural world, and disseminating and transforming ends and goals, as both an environment and way of life.
13. Sclove, R.E., *Democracy and Technology* (New York: Guildford Press, 1995).
14. *Ibid.*, Chapter 7.
15. *Ibid.*, p. 160.
16. Colin Ward, in *Housing: An Anarchist Approach* (London: Freedom Press, 1983), also made a similar case for participatory, local democracy in the development of urban neighbourhoods.
17. Milton Friedman, *Capitalism and Freedom* (University of Chicago Press, 1962), in his defence of "free market" globalisation, claimed that the globalisation of the market makes prejudice meaningless and obsolete. He quipped (p. 109) "The purchaser of bread does not know whether it was made from wheat grown by a white man or a Negro, by a Christian or a Jew." But, in my view, the problem is that when the process by which wheat is grown and made into bread is completely anonymous and abstract, the purchaser of bread simply does not care where it came from or how it was made. It becomes a product that comes into existence only at the moment that it is picked off the supermarket shelf. Globalisation generates an indifference to the material and social conditions for the production of our food. It allows us to be content to be ignorant about the conditions upon which our lives

are made possible, which is understandable, because once we are aware of the terrible conditions under which most of the people who grow our wheat in the Third World live and work, as well as the pittance that they are paid for their backbreaking efforts, then the bread becomes increasingly hard to swallow. Our only alternative to blissful ignorance is to develop callousness towards others as a precondition for our pleasure in our food.

18. *Democracy and Technology*, p. 85.
19. *Ibid.*, p. 204.
20. *Ibid.*, p. 104.
21. For further discussion of this point, see Harrison, P., *The Greening of Africa: Breaking Through the Battle for Land and Food* (London: Paladin Grafton Books, 1987); and Appfel-Marglin, F. and Marglin, S., eds, *Decolonising Knowledge: From Development to Dialogue* (Oxford: Clarendon Press, 1996).
22. *D&T*, p. 193.
23. See Dickson, D., *Alternative Technology and the Politics of Technological Change* (Collins, 1974) and *The New Politics of Science* (University of Chicago Press, 1984) for interesting and detailed discussion of the way that civic, military, and commercial ambitions have dominated the directions of science and technology since the end of the Second World War. See Frost, R.L., "Mechanical Dreams in Twentieth-Century France" (in Winner, ed., *Democracy in a Technological Society*, pp. 51–77), for a discussion of how French nuclear technology was developed as a symbol of national greatness, rather than simply a means of electricity supply. See also Myklebust, S., ed., *Technology and Democracy: Obstacles to Democratisation – Productivism and Technocracy* (Oslo: Centre for Technology and Culture, 1997); Postman, N., *Technopoly: The Surrender of Culture to Technology* (New York: Vintage Books, 1992); and Day, R.B., Beiner, R., and Masciulli, J., eds, *Democratic Theory and Technological Society* (Armonic, N.Y.: Sharpe, 1988), for further discussions, examples, and references.
24. See *The Whale and the Reactor*, Chapter 2; also, see *Autonomous Technology*.
25. *D&T*, p. 81.
26. *Ibid.*, p. 183.
27. Sassower, R., *Confronting Disaster: An Existential Approach to Technoscience* (Oxford: Lexington Books, 2004), pp. 72–3.
28. Winner, *Democracy in a Technological Society*, p. 15.
29. As C.P. Snow argued, in *Two Cultures and a Second Look* (Cambridge University Press, 1964), it is essential that we get beyond the "two cultures" science or humanities track education system. In *The Human Condition*, Hannah Arendt was critical of the ambition of technological science "to escape the human condition" instead of remaining philosophically related to it through a thinking relation with labour. Due to its increasing specialisation of technical and mathematical language, she argued that scientists and technologists increasingly "move in a world where speech has lost its power", where technology and thinking have "parted company for good" (p. 3).
30. *D&T*, p. 53.
31. See Ellul's discussion of this point in "Technology and Democracy" (in Winner, ed., *Democracy in a Technological Society*), pp. 43–4.
32. "Populism and the Cult of the Expert" (in Winner, ed., *Democracy in a Technological Society*, pp. 91–103).

33. *D&T*, p. 49.
34. *T&C*, p. 426.
35. *Ibid.*, p. 429.
36. Hughes, T.P., *Human-Built World: How To Think About Technology and Culture* (University of Chicago Press, 2004). See also Spirn, A., *The Granite Garden: Urban Nature and Human Design* (New York: Basic Books, 1984).
37. Hughes, *Human-Built World*, pp. 16–43. Miller, P., ed., *Errand into the Wilderness* (Cambridge, Mass.: Belknap Press, 1956); Thomas Jefferson, *Notes on the State of Virginia* (1785); Robert Beverly, *History and the Present State of Virginia* (1705); J.A. Etzler, *The Paradise Within the Reach of all Men, Without Labour, by the Powers of Nature and Machinery* (1836); and, Ralph Waldo Emerson, *Nature* (1836).
38. *HBW*, p. 153.

Index